MATHS
IN ACTION

Mathematics in Action Group

Members of the Mathematics in Action Group associated with this book:
D. Brown, R. D. Howat, E. C. K. Mullan, K. Nisbet, A. G. Robertson

**FURTHER
QUESTIONS**

Thomas Nelson and Sons Ltd
Nelson House Mayfield Road
Walton-on-Thames Surrey
KT12 5PL UK

Cover photograph by David Usill

First published by Blackie and Son Ltd 1986
New edition published by Thomas Nelson and Sons Ltd 1993

I⊤P Thomas Nelson is an International
       Thomson Publishing Company

I⊤P is used under licence

ISBN 0-17-431419-1
NPN 9 8 7 6 5 4

Printed in China

CONTENTS

INTRODUCTION

These *Further Questions* are intended to supplement the course developed in the series **Maths in Action**. They consist of exercises of harder questions, closely related to the corresponding (and similarly numbered) exercises in Book 1, and based on the text in Book 1. The 'F' notation, for example Exercise 1F, enables easy cross-reference to be made between these exercises and the A, B and C exercises in Book 1, especially where both are being used in the classroom or for homework.

1 WHOLE NUMBERS IN ACTION

EXERCISE 1F

At a golf driving range players are given 50 golf-balls to hit as far as they can. The scores are shown in the drawing. For example, 100–150 metres (including 100 but not 150) scores 2 points.

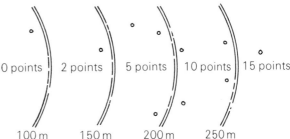

1 Paul has driven the ten golf-balls shown above. What is his score so far?

2 Julie and Nicholas each have a turn on the driving range. Here are their results:

	Distance in metres			
	100–149	150–199	200–249	250 and over
Julie	16	22	5	1
Nicholas	14	19	12	3

  a How many points did each person score?
  b How many non-scoring shots did each have?

3 Alastair played on his own and scored as follows:

Less than 100 m	100–149 m	150–199 m	200–249 m
1	16	23	7

  a How many times did he drive a ball 250 metres or further?
  b What was his score?

4 The highest score ever recorded at the driving range was 493. To reach this score the record-breaking player hit 14 shots of 250 or more metres, 23 shots between 200 and 249 metres, and 9 shots between 150 and 199 metres.
  a How many shots were hit a distance between 100 and 149 metres?
  b How many shots failed to travel 100 metres?

A ring road is built around a city. In the drawing, distances are in kilometres and motorway junctions are numbered ① to ⑦.

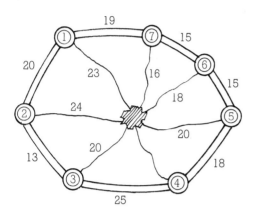

5 How far is it around the motorway?

6 How far is it from junction ② to junction ⑤:
  a through the city
  b on the motorway, travelling
     (i) clockwise   (ii) anticlockwise?

7 The distance from junction ① to junction ④ through the city is 10 km shorter than the shortest distance by motorway. How far is it from the city to junction ④?

8 Mr Walters has to check the emergency telephones at junctions ①, ③, ⑤ and ⑦, in that order. What is the shortest distance he must travel?

EXERCISE 2F

1 Copy and complete these magic squares:

a

14		12
	15	
		16

b

		31
32	37	36

c

	4		
	6		3
14	7	11	2
	1	8	

2 Make up a '3 by 3' magic square of your own, using the numbers 21, 22, 23, ..., 29.
Hint Your answer to question **1a** could help.

3 Copy and complete:

a

7			11	
13	21			5
19		15		6
25	8	16	4	
1		22	10	

b

36			4		35	31
		29	27	10	26	
19	17			21	14	18
13			16	15	23	24
		11	9		8	25
6	32		34			1

4 A calculator, and some friends, would help you to complete:

1	99	3			5	94	8	92	10
90				86	85	17	83	19	11
80	79	23	77		26	74	28	22	
31	69	68	34	66	65	37	33	62	
60		58		45	46	44	53	49	51
	52		47	55	56		48		41
61	32		64	36		67	63	39	70
21	29	73	27	75		24	78	72	30
20	82	18	84	15	16	87	13	89	
91	9	93	4	6	95	7	98		100

EXERCISE 3F

---

1 Which pairs of numbers are equal?

| ONE MILLION | TEN THOUSANDS | ONE THOUSAND THOUSANDS |

| TEN HUNDREDS | ONE HUNDRED HUNDREDS | ONE THOUSAND |

2 Copy and complete the calculations, but don't use the number 1 on its own. For example, $12 \div \square \times \square = 8$ becomes $12 \div 3 \times 2 = 8$.

a $15 \div \square \times \square = 35$ **b** $28 \div \square \times \square = 21$ **c** $27 \div \square \times \square = 24$

d $36 \div \square \times \square = 20$ **e** $48 \div \square \times \square = 6$ **f** $56 \div \square \times \square = 42$ (two answers)

3 Arrange the number cards in two equal groups, each with a total of 18. For example, $7 + 1 + 4 + 6 = 18$ and $2 + 3 + 5 + 8 = 18$. How many solutions can you find?

4 The 'In' and 'Out' number is the same. Find the number for each calculation.

a

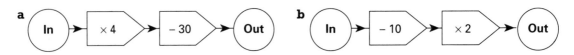

b

5 Using each number 1, 2, 3 once, and the operations + and ×, you can obtain 5, 6, or 7. For example, remembering '× before +':
$2 + 1 + 3 = 6$, $1 + 2 \times 3 = 7$ (check this), $1 \times 3 + 2 = 5$, $1 \times 2 \times 3 = 6$.
  a Can you find eight different answers, using 1, 2, 3, 4 in the same way?
  b Using brackets, more numbers can be found. For example, $(3 + 1) \times 2 = 8$.
    (i) Use 1, 2, 3, 4, +, ×, and ( ) to make all the numbers from 9 to 21.
    (ii) What is the largest number you can find?

EXERCISE 4F

1 Write down two more terms for each sequence, and describe the rules you use.
  a 1, 8, 15, ...   **b** 8, 17, 26, ...   **c** 1, 2, 5, 10, 17, ...
  d 100, 90, 81, 73, ...   **e** 1, 3, 9, ...   **f** 2, 3, 5, 6, 8, ...

2 Make as many sequences as you can which start 1, 4, ... Describe the rule for each one.

3a Copy and continue each row in this pattern of numbers for three more numbers in each row.

$$\begin{array}{ccccccc} 1 & & 4 & & 9 & & 16 \\ & 3 & & 5 & & 7 & \cdots \\ & & 2 & & 2 & & \cdots \end{array}$$

  b Explain how to obtain each row from the row above it.

4a Continue this dot pattern for two more terms.

  b Find the fourth, fifth, tenth and hundredth term in each of these sequences:
    (i) 1×3, 2×4, 3×5, ...   (ii) 1×2, 2×4, 3×6, ...

5 Find the sum of each of the following mentally:
  a $1 + 2 + 3 + 4 + 5$ **b** $1 + 2 + 3 + 4 + 5 + 6 + 7 + 8 + 9 + 10$ **c** $1 + 2 + 3 + ... + 98 + 99 + 100$

EXERCISE 5F

No calculators in this exercise!

1 Calculate:

a 10×50 **b** 25×0 **c** 100×33 **d** 40×1000

e 1000×1000 **f** $1 \times 10 \times 100 \times 1000$ **g** $10 \times 10 \times 10 \times 10$ **h** $10\,000 \times 1000$

2 Calculate:

a 32×5 **b** 106×4 **c** 6×75 **d** 8×505

e 0×318 **f** 20×20 **g** 34×40 **h** 50×700

i 700×700 **j** 8000×300 **k** 900×84 **l** 200×896

3 Find these products:

a 1×77 **b** 99×3 **c** 19×30 **d** 101×50

e 999×0 **f** 400×25 **g** 1111×60 **h** 700×203

i 125×800 **j** 6000×43 **k** 11×110 **l** 101×11

4 Calculate:

a $3 \times 100 \times 3$ **b** $5 \times 17 \times 4$ **c** $1000 \times 6 \times 12$ **d** $99 \times 10 \times 5$

e $8 \times 125 \times 100$ **f** $60 \times 60 \times 60$ **g** $2 \times 74 \times 50$ **h** $10 \times 9 \times 99$

i $80 \times 16 \times 50$ **j** $100 \times 90 \times 80$ **k** $300 \times 300 \times 300$ **l** $10\,000 \times 1000 \times 10 \times 0$

EXERCISE 6F

1 London

London				
1550	Moscow			
10 565	9425	Sydney		
6218	4668	4640	Tokyo	
3672	4884	9792	6763	Washington

The table shows distances by air between some cities.

a Find the distance, to the nearest thousand miles, from:
(i) London to Tokyo (ii) Tokyo to Moscow.

b Estimate the distance from London to Tokyo to Moscow, to the nearest thousand miles.

c Use the table to calculate the actual distance from Tokyo to Moscow.

d Repeat **a**, **b** and **c** for the journey London to Sydney to Washington.

2 The table shows the average daily sales of four newspapers. Calculate:

a the entries for (i), (ii), (iii) and (iv)

b the total daily sales of the four papers in (i) 1993 (ii) 1992.

	1993	1992	change '92–'93
Daily Globe	429 794	426 331	(i)
The Review	(ii)	597 362	Up 17 608
Top News	980 391	(iii)	Down 31 214
The Editor	(iv)	724 683	Down 41 217

3

Number of matches	40	41	42	43	44	45
Number of boxes	2	5	8	10	5	1
Total number of matches	80					

a How many boxes of matches contain:
(i) 42 matches (ii) more than 42 matches?

b Calculate: (i) the entries in the bottom row (ii) the total number of matches.

4 Mrs Webb spends 60p daily on newspapers from Monday to Saturday. On Sunday she spends £1.30. Calculate her bill for the week, and her average daily bill.

5 The table shows the number of goals scored in a football league one Saturday.

Number of goals	0	1	2	3	4	5	6
Number of teams	7	10	10	6	4	2	1

Calculate: **a** the total number of: (i) teams (ii) goals scored.
b the average number of goals scored per team.

2 ANGLES AROUND US

EXERCISE 1F

1a Fold a piece of paper across the middle. Call the crease ABC.
  b Fold the paper again so that the crease AB lies along BC.
  c How do you know that: (i) ∠ABC = 180° (ii) ∠ABD = 90°?

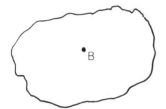 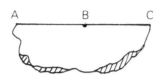

2a Use your right-angled piece of paper to find out which of these angles are acute and which are obtuse.

(i)    (ii)    (iii)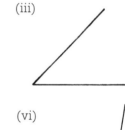

(iv)    (v)    (vi)

  b By making one more fold in your paper, measure the sizes of two of the angles, without using a protractor.

3a Find the maximum number of obtuse angles you can have in:
    (i) a triangle (ii) a 4-sided figure (iii) a 5-sided figure.
  b Draw an example of each.

4 Peter decided that there are only three types of triangle, based on the kinds of angles they have. He made a list. Explain what it means, and draw a triangle of each type.
  A A A, A A Rt, A A O

ACUTE
OBTUSE
RIGHT
REFLEX

5 Sasha is investigating the different types of quadrilateral in the same way. Her list looks like this:
Rt Rt Rt Rt, Rt Rt O A, Rt O A A.
Draw her quadrilaterals. Can you find more types?

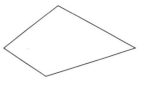

Quadrilateral

6

No. of sides	3	4	5	6
Max. no. of rt. ∠s	1	4	3	5

Alan was investigating the maximum number of right angles in shapes with different numbers of sides.

a Draw examples of the four shapes he discovered.
b Investigate shapes with more sides to find evidence for a pattern of entries in the table.

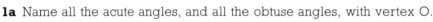

EXERCISE 2F

The Yellow Darts air display team 'burst' from point O.

1a Name all the acute angles, and all the obtuse angles, with vertex O.
b Name all the acute angles, and all the obtuse angles, with OA as one arm.
c Name all the reflex angles.

2 A snooker ball rebounds from the cushion at the same angle as that at which it strikes it. Calculate a, b, . . .

(i)

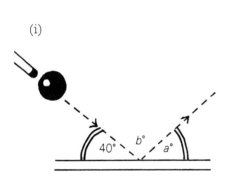

(ii)

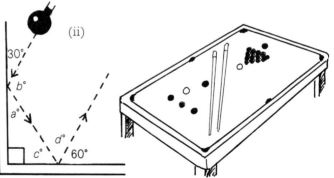

3 Angle u is 15 times angle v. If 90 of angle v fit round a point, how many of angle u fit round a point?

4a Using tracing paper, or another method, find how many of angle x and how many of angle y are needed to fit around a point.
b How many times greater is y than x?

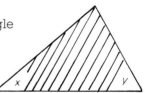

5 There are two types of set-square.
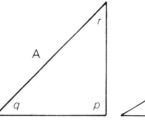
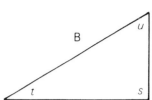

   a How many set-squares of type A would
    you need for:
    (i) p (ii) q (iii) r to fit around a point?

   b How are angles p, q, r connected to
    each other?

   c Investigate the type B set-square angles
    in the same way.

6 In a diagram, reflex $\angle AOB = 210°$ and reflex $\angle COB = 260°$. Calculate the size of $\angle AOC$.
Is there more than one answer?

7 These diagrams show which floor the lift is at in each hotel.

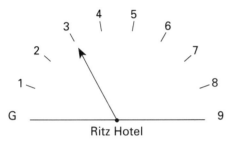
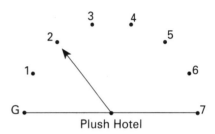

   a Calculate the angle turned through by the arrow in the Ritz Hotel when the lift travels
    from floor:
    (i) 3 to 7 (ii) 8 to 5 (iii) 2 to 9 (iv) 8 to 2.

   b Repeat **a** for the Plush Hotel (giving angles to the nearest degree) for travel from floor:
    (i) 2 to 6 (ii) 3 to 5 (iii) 7 to 2 (iv) 1 to 7

8 The spokes on these wheels are equally
spaced.

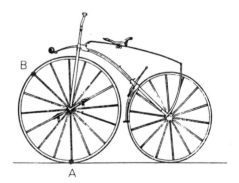

   a The bicycle has moved forward until B
    has touched the ground. Through what
    angle has the wheel turned?

   b During the movement in **a** the back
    wheel has turned through 180°. How
    many turns has the back wheel made if
    the front wheel has made three turns?

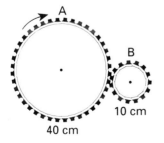

9 Gearwheel A turns gearwheel B. The circumferences are
40 cm and 10 cm. As A turns once, B turns 4 times.
Complete:

   a As A turns 90°, B turns . . .

   b B turns 180° because A turns . . .

   c As B makes 2 turns, A makes . . .

   d As A makes 2 turns, B makes . . .

EXERCISE 3F

1 How many times greater is angle *z* than: **a** angle *x*   **b** angle *y*?

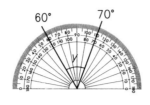

2a Write down the size of:
   (i) ∠BAC (ii) ∠ABC.
b Measure the size of
   ∠ACB.

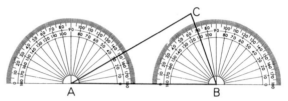

3

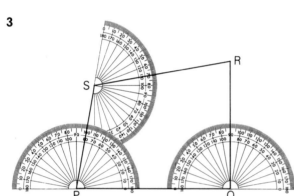

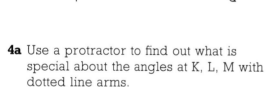

a Write down the sizes
   of angles
   (i) PQR
   (ii) QPS
   (iii) PSR (be careful).
b Measure the size of
   ∠QRS.

4a Use a protractor to find out what is
   special about the angles at K, L, M with
   dotted line arms.
b Draw a triangle, and the lines bisecting
   the angles. Do the lines meet at a point?

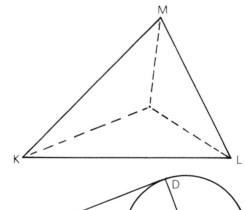

5a Use a ruler and protractor to find pairs
   of equal lines and angles in the diagram.
b Calculate ∠BAD + ∠BCD.

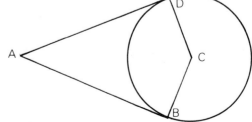

EXERCISE 4F

1a Draw a line AB 5 cm long. Moving anticlockwise, draw:
(i) ∠BAC = 30°   (ii) ∠CAD = 110°   (iii) ∠DAE = 40°.
b Check that EAB is a straight line. Why is this?

2 Use the flowchart to draw each trapezium, then copy and complete the table:

x	u	y	v	u + v
5	40	2		
6	90	3		
7	125	4		

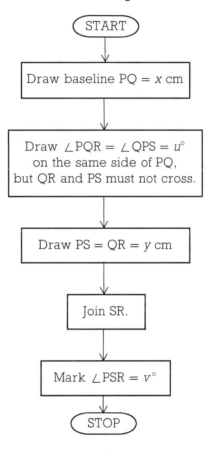

To draw a trapezium

START

Draw baseline PQ = x cm

Draw ∠PQR = ∠QPS = u°
on the same side of PQ,
but QR and PS must not cross.

Draw PS = QR = y cm

Join SR.

Mark ∠PSR = v°

STOP

3 This arrow is balanced about its central line, so you should have enough information to make an accurate drawing of it. Try to do this.

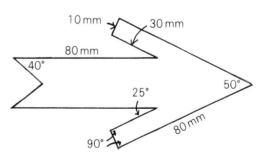

EXERCISE 5F

1 In (i) the cube is placed on its base, in (ii) on an edge and in (iii) on a corner.

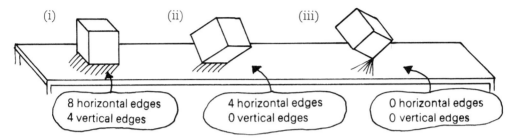

(i) (ii) (iii)

8 horizontal edges
4 vertical edges

4 horizontal edges
0 vertical edges

0 horizontal edges
0 vertical edges

No matter how the cube is moved about, these are the only three possibilities for the number of horizontal and vertical edges.

Investigate the number of possibilities in the same way for **a**, **b** and **c** and then for a solid of your own choice.

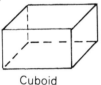
a
Cuboid

b
Tetrahedron

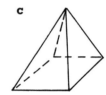
c
Pyramid on square base

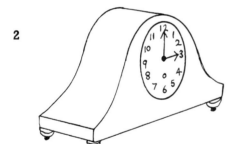

2 The clock-face is vertical. Investigate when the two hands are:
a both vertical
b both horizontal
c perpendicular (at which hours, and how often between hours).

3 Investigate parallel and perpendicular edges for a cube, cuboid, tetrahedron and pyramid on a square base.

4a This wallpaper has a pattern of squares. How many sides of the four squares you can see in the picture are vertical, and how many are horizontal?
b Each square is replaced by an equilateral (equal sided) triangle, and the same questions are asked. How many different answers are possible? Explain.

A marble, not rolling

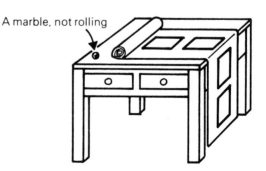

3 LETTERS AND NUMBERS

EXERCISE 1F

This is a 1 kg weight:

The weights come in these containers:

Labels on the containers give the number of weights inside.

Cans weigh 5 kg, empty

Boxes weigh 3 kg, empty

Bags weigh 1 kg, empty

1 Readings on the weighing machines are in kg. What number does each letter stand for?

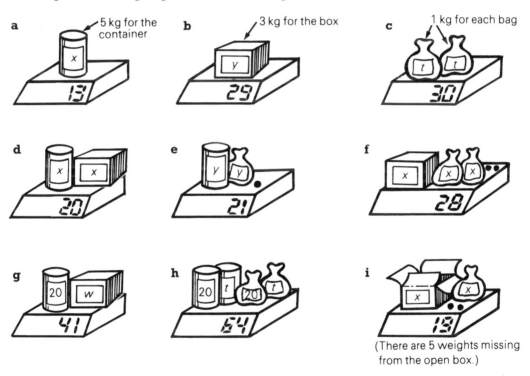

a ← 5 kg for the container

b ← 3 kg for the box

c ← 1 kg for each bag

i (There are 5 weights missing from the open box.)

Use the information in *both* pictures below to find the values of *x* and *y*.

j

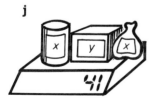 

2 All these weighing machines are balanced. Find the value of the letter in each case.
(Remember the weights of the containers!)

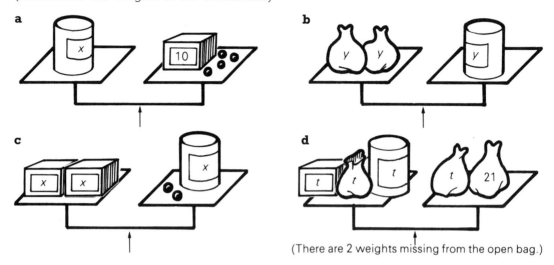

(There are 2 weights missing from the open bag.)

3 What number did each letter replace in these spreadsheets?
Give your answers in the form $a = \ldots$

a

	A	B	C	D	E	F	G	H	I
1	1	1	2	3	5	h	13	21	j
2	2	3	5	7	10	14	20	29	g
3	a	5	8	11	15	20			
4	4	c	11	15	20	26			
5	5	9	14	d	25				
6	6	11	17	23	30				
7	b	13	20	e					
8	8	15	23	f					

b

	A	B	C	D	E
1	1	4	9	16	25
2	2	7	16	29	46
3	x	13	30	t	a
4	8	25	w	107	b
5	16	y	114	r	c
6	32	97	e	419	676

EXERCISE 2F

1

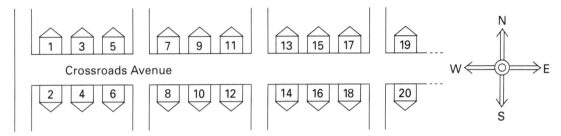

a Mr Newson lives at house number $x + 3$ on the South side of Crossroads Avenue. What is the number of the house:
   (i) opposite
   (ii) next door to the East
   (iii) next door the West
   (iv) 5 houses to the East
   (v) 2 houses to the West
   (vi) opposite his neighbour to the East?

b Ms Ayers lives at a corner house, number y, Crossroads Avenue. Her friend Ms Cross lives diagonally opposite in another corner house.
   (i) What are the possible numbers for Ms Cross's House?
   (ii) If Ms Ayers' house is an odd number and Ms Cross's number is smaller than hers then which of the possibilities is correct?

c Mr Lord lives at number $x - 2$ and Mr Christianson at number $x + 8$ Crossroads Avenue. Give instructions to Mr Lord to get to Mr Christianson's house from his own.

2

This 12 page pamphlet is made from three sheets of paper:

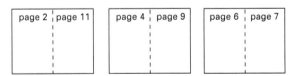

a Turn each sheet over and show the page numbering on the other side.

b

What page number is this?

c Turn this sheet over and show the page numbering on the other side.
d Repeat **b** and **c** for a 20 page booklet.

3

Spreadsheet I

	A	B	C	D	E	F	G
1	20	23	26	29	32	35	38
2	19	22	25	28	31	34	37
3	18	21	24	27	30	33	36
4	17	20	23	26	29	32	35
5	16	19	22	25	28	31	34
6	15	18	21	24	27	30	33

Spreadsheet II

	A	B	C	D	E	F	G
1	20	21	22	23	24	25	26
2	17	18	19	20	21	22	23
3	14	15	16	17	18	19	20
4	11	12	13	14	15	16	17
5	8	9	10	11	12	13	14
6	5	6	7	8	9	10	11

Spreadsheet III

	A	B	C	D	E	F	G
1	20	21	22	23	24	25	26
2	22	23	24	25	26	27	28
3	24	25	26	27	28	29	30
4	26	27	28	29	30	31	32
5	28	29	30	31	32	33	34
6	30	31	32	33	34	35	36

Spreadsheet IV

	A	B	C	D	E	F	G
1	20	17	14	11	8	5	2
2	23	20	17	14	11	8	5
3	26	23	20	17	14	11	8
4	29	26	23	20	17	14	11
5	32	29	26	23	20	17	14
6	35	32	29	26	23	20	17

Spreadsheet V

	A	B	C	D	E	F	G
1	20	22	24	26	28	30	32
2	19	21	23	25	27	29	31
3	18	20	22	24	26	28	30
4	17	19	21	23	25	27	29
5	16	18	20	22	24	26	28
6	15	17	19	21	23	25	27

Spreadsheet VI

	A	B	C	D	E	F	G
1	20	18	16	14	12	10	8
2	21	19	17	15	13	11	9
3	22	20	18	16	14	12	10
4	23	21	19	17	15	13	11
5	24	22	20	18	16	14	12
6	25	23	21	19	17	15	13

a Which spreadsheets are these from? Give reasons for your answers.
Careful! There may be more than one possibility.

(i)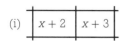

(ii)

(iii)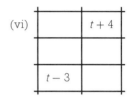

(iv)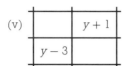

(v)

(vi)

b Identify the correct spreadsheet, and find the value of the letter in each case:

(i)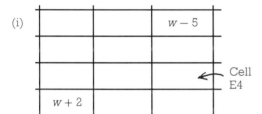

(ii)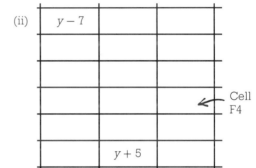

(iii)

EXERCISE 3F

1a Julie opens a bank account with £x on Monday. Write down her 'balance' each day.

Day	Action	Balance (£)
Monday	Puts in £x	x
Tuesday	Puts in £5	
Wednesday	Takes out £3	
Thursday	Puts in £7	
Friday	Takes out £9	
Saturday	Puts in £6	

b Jamie opens an account and also puts in £x on Monday. Complete this balance sheet.

Day	Action	Balance (£)
Monday	£x in	x
Tuesday	£x in	
Wednesday	£7 in	
Thursday	£x in	
Friday	£(x + 7) out	
Saturday	£(x + 1) in	

2 Write in shorter form:

a $t + 1 + t + 1$ **b** $7 + 4t + 3$ **c** $8 + t - 6 + t$ **d** $2 + 2t + 2t - 2$

e $2t + 2 + t$ **f** $2t + 5 + 2t + 5$ **g** $t + 7 + 3 + t - 8$ **h** $3t + 1 + t - 1$

You should have four pairs of equal answers. Which pairs?

3   *Ann:*   'Add $3m$, $3m$, 2 and 2'

   Mary: 'Multiply m by 4, add $2m$, add 8'

     Ian: 'Multiply m by 2, add four times m, subtract 4'

     Sue: 'Multiply m by 3, subtract 2, add $3m$, subtract 2'

Simplify each answer. Which ones are the same?

4a Copy and complete these spreadsheets where each *row* has the given rules (A1, A2, A3 and A4 are the input cells).

(i)

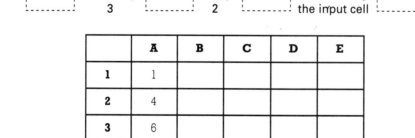

	A	B	C	D	E
1	1				
2	4				
3	6				
4	9				

(ii) Input cell   Output cell

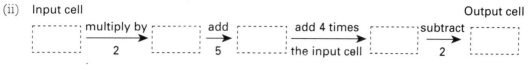

	A	B	C	D	E
1	1				
2	4				
3	6				
4	9				

b For each of these two spreadsheets, put x in the input cell and find the expression for the output cell. Compare the two output cells and comment.

5a Copy and complete these two spreadsheets:

(i)

	A	B	C
1	Number of	Cost of	Total cost
2	items:	each (£):	(£):
3	4	x	$4x$
4	6	3	
5	3	x	
6	2	4	
7			

(ii)

	A	B	C
1	Number of	Cost of	Total cost
2	items:	each (£):	(£):
3	4	3	
4	5	x	
5	4	4	
6	1	x	
7			

b In each spreadsheet, cell C7 gives C3 + C4 + C5 + C6. Find the expression in cell C7 in each case.

c The numbers that appear in cell C7 are the same on the two spreadsheets. What is the value of x?

4 DECIMALS IN ACTION

EXERCISE 1, 2F

---

1 In the number 98 765.432, 9 has a place value of 90 000.
Write down the place value of:
a 7 **b** 6 **c** 4 **d** 2

2 Arrange these numbers in order, starting with the largest: 12.10, 9.79, 9.97, 12.01, 10.05.

3 Explain the value of each digit in these atomic weights:
a copper, 63.546 **b** aluminium, 26.98154

4 Write down the numbers on the rulers which the arrows point to:

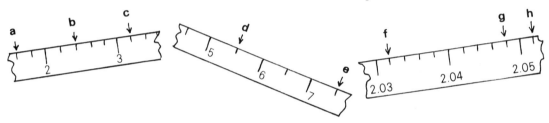

5 A measuring jug is marked every tenth of its volume.
Draw a sketch of the jug when it is:
a 0.1 full **b** 0.5 full.

6 Write down the readings on the pressure gauge indicated by the pointers.

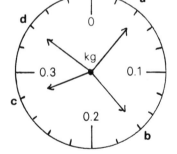

7 On Sunset Island the hunters only have one hand free at a time. In the other hand they hold their blowpipe. So their counting system is 0, 1, 2, 3, 4, 10, 11, 12, 13, 14, 20, 21 . . .
a Write down their next six numbers.
b Ipo writes down 12. How do we write this number?
c A visitor offers them 21 beads for a blowpipe. How many is this in their system?

8 Sue tried to crack the bar code in her shop.

a How does the code work?

b How much is this item?

c What is the code for 45p?

EXERCISE 3, 4F

1a Add 4.81 to 17.6 and subtract 6.08 from the sum.
 b Calculate $14.8 - 2.91 - 8.9 + 0.78$.

2 Sheena's time on the downhill ski race is 2 minutes 53.28 seconds. Astrid is three tenths of a second faster, but Leila is 6 hundredths of a second slower. What are their times?

3 The school's new music block is built in the shape of a square with a square garden in the centre. Calculate, in metres,
 a the distances around the outside and inside edges of the building
 b the difference between these distances.

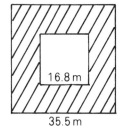

16.8 m

35.5 m

4 Clare empties her basket at the checkout and the till registers: 'Fresh orange 128p, Highland Sparkle 68p, margarine 126p, ice-cream surprise 87p and budgie seed 129p less 10p special offer.' How much change does Clare receive from her £10 note?

5 The lap times in the sports day relay race are 53.8, 54.7, 52.9 and 54.4 seconds. Calculate:
 a the total time for the race
 b the difference between the fastest and slowest laps.

6 Three pupils in the home economics class are measuring lengths of cloth A–G, in metres. Calculate:
 a the total length of cloth measured by each pupil
 b the difference in the totals between the pupils
 c which material seems to be the most difficult to measure.

Material	A	B	C	D	E	F	G
Charles	2.6	1.6	4.9	3.9	1.9	4.5	0.6
Helen	2.5	1.7	4.7	4.0	2.0	4.2	0.5
Pat	2.55	1.55	4.75	3.95	1.95	4.05	0.45

7 The Walker family are off to America. The total weight of their four cases is 83.78 kg, and the Airline's maximum weight per case is 25 kg. Is the Walkers' fourth case overweight?

24.35 kg

17.08 kg

14.67 kg

EXERCISE 5, 6F

1 List the totals for these order books.

Number ordered	Length (m)	Total (m)
10	5.82	
100	4.96	
1000	1.06	

a In the joiner's shop.

Number of samples	Weight (g)	Total (g)
10	0.41	
100	0.09	
1000	1.25	
10·000	0.085	

b In the laboratory.

2

£2.65   £5.85   £12.70   £48.65

Calculate the cost of these calculators:
a 20 graphic   **b** 60 graphic   **c** 40 programmable   **d** 200 programmable
e 50 scientific   **f** 300 scientific   **g** 80 basic   **h** 2000 basic

3 'Yes, 38 compact discs at £12.96. About £520', thought Eve. How did she work out her answer? Calculate the exact answer.

4 Tim was making a rough check of the value of stock on the shelves. Complete it for him.

48 TINS OF BEANS AT 39p EACH =50 ×... = ...   18 PKTS OF CEREAL AT 108p EACH= . . . . .
95 TINS OF SOUP AT 42p EACH= . . . .   31 JARS OF COFFEE AT £1.16 EACH= . . . .

5 Calculate the exact costs in question **4**, also the total value.

6 In each of the following, write down an approximation and *then* calculate the exact value.
  a 1.96 × 38   **b** 183.7 × 42   **c** 2.039 × 18   **d** 101.9 × 101.

7 Before setting off for a holiday in Germany, Ellen changed £180 to marks, at 2.855 marks to the £. How many marks was she given?

8 On holiday at the seaside, Jenny lay awake counting the number of times the lighthouse beam lit up her room. The beam passed every 13.5 seconds and after counting 25 Jenny fell asleep. How long, in minutes and seconds, did she lie awake?

9 Mr Brown had £20 000 redundancy money. He invested it in a monthly income account which gave him 12.75p interest on every £1 each year. How much income had he after a year and how much after income tax at 25p in the £ was taken off?

10 This is part of Jean's itemised telephone bill:
   a Calculate the total time and the total cost.
   b What is the average cost per minute?

Duration (min)	Charge (£)
9	1.050
7	2.390
5	1.695
6	1.695
3	1.000

EXERCISE 7F

1 Round these numbers to 3 decimal places:
   a 3.4676   **b** 12.0608   **c** 1.2355   **d** 0.0474

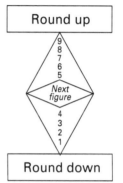

Round up

Next figure

Round down

2 Round to the nearest penny:
   a £8.379   **b** £2.015   **c** £1.096   **d** £0.813.

3 Round these measurements:
   a 20.9 days to the nearest day
   b 31.29 s to the nearest second
   c 36.75° to the nearest tenth of a degree
   d 4.468 tonnes to the nearest tonne
   e 14.385 km to the nearest hundredth of a kilometre
   f 1.7809 kg to the nearest thousandth of a kilogram.

4 Wecutem Co supplies rectangular metal sheets, cut to size. A customer requires sheets measuring 1.6 m by 2.3 m, correct to 1 decimal place. Which of these sheets would be acceptable?
   a 1.64 m by 2.28 m   **b** 1.57 m by 2.36 m   **c** 1.52 m by 2.09 m   **d** 1.59 m by 2.30 m

5 Four numbers have been rounded from 3 decimal places to 2 decimal places to give the numbers below. Write down the largest and smallest possible values of the original four numbers.
   a 15.36   **b** 0.08   **c** 7.03   **d** 34.79

6 When £20 is shared equally among seven friends, how much does each receive, and how much is left over?

7 Ella receives £25.75 for 6 hours work and Kris gets £31.50 for 7 hours work. Who has the higher hourly rate and by how much, to the nearest penny per hour?

8 $24 \times n = 201.6$. Find number n.

9 Sixty-five tickets are printed for a '12–16 club' disco. The cost of running the disco is £128.
   a What is the least price that has to be charged for a ticket so that a loss is not made?
   b What is a ticket likely to cost?

10 $27.3 \times n = 379.47$. Find number n.

11 A bottle of concentrated orange juice contains 1.65 grams of vitamin C. How many bottles are needed to provide at least 10 grams of vitamin C?

12 Red Delicious apples cost £1.72 per kg. What is the weight of a bag of apples costing £2.15?

13 Sharon gets £7.50 for $3\frac{1}{2}$ hours babysitting. Her brother Jim is paid £4.65 for $2\frac{1}{4}$ hours spent delivering leaflets. Which task is better paid?

14 Packets of salt weighing 1.8 kg are filled from a 50 kg sack.
   a How many packets can be filled?   **b** What is the weight of salt left over?

15 Mrs Fox uses 9.7 gallons of petrol on a car journey where the mileometer changes from 27 635 miles to 28 063 miles. Calculate the average petrol consumption in miles per gallon to the nearest whole number.

5 FACTS, FIGURES AND GRAPHS

EXERCISE 1F

1 Use the playing cards shown to help you copy and complete the tables below.

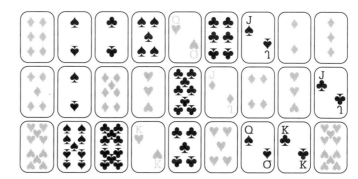

a

	Tally	Number
Red		
Black		

b

Suit	Tally	Number
Hearts		
Clubs		
Diamonds		
Spades		

c

Value	Tally	Number
A to 4		
5 to 9		
10 to K		

2 The number of children in each of the families of pupils in 1G is:
2, 1, 4, 1, 3, 3, 2, 2, 3, 5, 4, 2, 2, 5, 1, 7, 4, 3, 2, 3, 1, 4, 2, 3, 2, 1, 2, 2

   a Copy and complete the table.
   b How many pupils are in 1G?
   c Which family size is most common?

No. of children	1	2	3	4	5 or more
Tally					
No. of families					

3 Mr Wood plays in a golf competition every weekend. A score-card reading 76/8 means he scored 76 and came 8th in the competition.

   a Copy and complete the tally charts for his rounds of golf:
   76/8, 69/2, 70/4, 67/1, 76/9, 75/7, 71/4, 74/7, 72/3, 68/3, 73/5, 71/3, 74/6, 73/4, 68/3, 74/1, 72/5, 75/8, 70/3, 69/2, 74/5, 70/4, 75/8, 74/7, 72/5, 71/3, 73/3, 76/7.

Score	67	68
Tally		
Total		

Position	1	2
Tally		
Total		

   b What score, and what position did he achieve most often?

4 Kath asks pupils in her class how long they take to travel to school. The times, in minutes, are:
16, 22, 8, 17, 35, 28, 11, 31, 25, 18, 13, 15, 20, 16, 23, 31, 7, 19, 20, 13, 10, 34, 18, 25, 11, 26, 22, 19, 20.

a Copy and complete her table of times.
b Why does she group the times?
c Which group has most pupils?

Time	6–10	11–15	16–20	21–25	26–30	31–35
Tally						
Total						

5

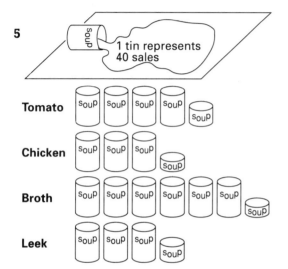

1 tin represents 40 sales

The supermarket recorded its weekly sales of soups. This pictogram appeared in their report.

a What was the most popular brand of soup?
b How many more tomato did they sell than chicken?
c If they originally had 300 cans of each type of soup, how many of each are now left in stock?
d Represent the stock of soups now held as a pictogram.

6 The table shows the number of goals scored in the league on 10th May.

Div. 1	Div. 2	Div. 3	Div. 4
14	18	25	23

a Draw a pictogram, using
⊕ = four goals

b What was the total number of goals? Show this total, using the same scale.

7 The pictogram shows the number of trees, to the nearest hundred, felled by Tall Timber Company in a four year period.

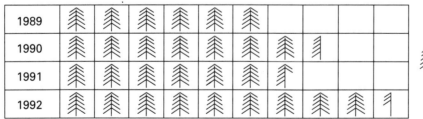

= 1000 trees

a In which year was the greatest number of trees felled?
b Between which years did the number of trees cut down decrease?
c List the years in the order of *the number of trees felled*, largest number first, and give the number felled each year.
d In 1993, 12 697 trees were cut down. Show this, using the 1000 tree symbol.

EXERCISE 2F

1

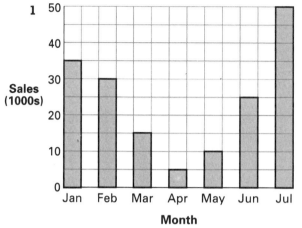

The sales of a magazine were starting to fall. An advertising campaign was carried out. The chart indicates the sales over the period of the campaign.

a What were the sales in February?

b In which month did the campaign begin to have an effect?

c What were the total sales for the two months: (i) before the month in **b**, (ii) after the month in **b**?

2 The bar chart shows six weeks' sales of the latest record by *The Magic Squares*.

a Describe the pattern of sales.

b (i) In which week did the sales reach their peak?

(ii) How many were sold that week?

c Between which two weeks was the greatest drop in sales?

d Calculate the total sales for the six week period.

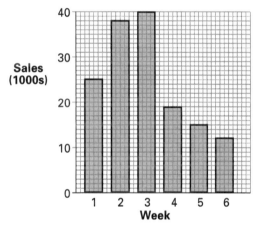

3

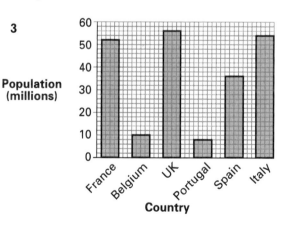

Sandra drew a bar graph of the population of some European countries.

a List the countries and their populations, largest first.

b The population of Norway is 8 million. How many squares high should its bar be drawn?

4 The table shows the heights, to the nearest 100 feet, of the highest mountains in the British Isles.

Ben Nevis	Scafell Pike	Snowdon	Carrountoohil
4400	3200	3600	3400

Draw a bar graph of the information.

5

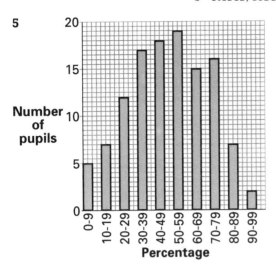

The exam marks of first year pupils have been organised in the form of a bar graph.
a Why do you think they have been grouped in class intervals?
b In which class do most pupils have their mark?
c How many pupils scored more than 59?
d How many pupils are in the first year?

6

Prices well within your pocket.

£1.23	£0.37	£2.45	£0.35	£3.29
£0.89	£4.98	£4.28	£2.29	£2.67
£3.45	£1.65	£1.54	£1.87	£1.29
£4.21	£2.43	£0.75	£0.55	£3.00
£4.20	£0.10	£0.60	£1.40	£2.10
£4.65	£2.00	£1.40	£3.40	£1.00
£0.05	£0.99	£1.95	£0.60	£1.00
£2.24	£0.56	£1.90	£2.00	£0.35

Bill saw the claim in the shop. He took a note of forty different prices.
a Arrange the data in suitable classes.
b Draw a bar graph to illustrate your figures.
c Comment on the shop's claim.

7 A very special number in mathematics is denoted by the Greek letter π (pi). Written out as a decimal, this number goes on forever. In the card opposite it has been given to 170 decimal places. Many mathematicians have looked for patterns in these numbers.
Draw a bar graph to show the occurrence of the digits in π.

π = 3.14159 26535 89793
23846 26433 83279 50288
41971 69399 37510 58209
74944 59230 78164 06286
20899 86280 34825 34211
70679 82148 08651 32823
06647 09384 46095 50582
23172 53594 08128 48111
74502 84102 70193

8

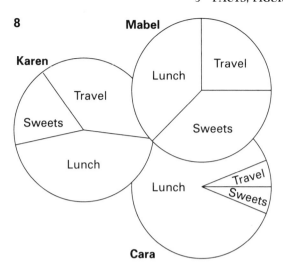

Karen, Mabel, and Cara compare the ways in which they spend their pocket money on a typical school day.

a Who spends least on travel?

b Mabel spends £3.00 a day; Karen £3.60 and Cara £4.00. With the aid of the pie charts, estimate how much each spends on sweets.

c Fill in the following spending table, using reasonable estimates.

	Lunch	Sweets	Travel
Mabel			
Karen			
Cara			

9 The Maths Department has £3600 to spend.
The angles at the centre of the pie chart are shown.

a What is most money spent on?

b How much is spent on calculators?

c How much more is spent on paper than on jotters?

d *Pencils* is used as a label to describe all drawing implements.
How much is spent on this item?

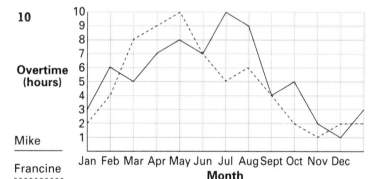

10

Overtime (hours)

Mike ——————

Francine --------

Francine and Mike work for the same firm but have different patterns of overtime during the year.

a In which months was Francine's overtime more than Mike's?

b In which months was their overtime the same?

c How much overtime did each work from January to December?

11

Electric	Gas
95°C	1/4
110°C	
130°C	1/2
140°C	1
150°C	2
160°C	3
180°C	4
190°C	5
200°C	6
220°C	7
230°C	8
240°C	9

Using the data in this table, draw a line graph which illustrates the relation between the setting required when using an electric cooker (°C) and when using a gas cooker (gas mark).

Use your graph to answer the following:

a What gas setting is equivalent to 170°C?

b What electric setting is the same as gas mark 3.25?

c The gas is turned up from gas mark 2.5 to 7.25. How many °C rise in temperature is this?

6 MEASURING TIME AND TEMPERATURE

EXERCISE 1F

1a Calculate the number of days, including both dates, from 25th March to 3rd November.
b The 15th of May is a Tuesday. What day of the week is the 20th of June?

2 Write down the numbers that go in these spaces:
  a ... seconds in an hour   **b** ... minutes in a day
  c ... seconds in a day   **d** ... hours in a week
  e ... minutes in a week   **f** ... hours in a year (two answers).

3 a Around the world in 80 days! If you set out on Tuesday, on what day would you return?

  b If March 3rd is a Monday, what day of the week could March 3rd be in the following year?

4 The nineteenth century is one century ago. How many centuries ago is:
  a the seventeenth century AD   **b** the first century AD
  c the first century BC   **d** the fifth century BC?

5 a How many leap years have you lived through?
  b So, how many:
   (i) days
   (ii) hours
      old are you?

EXERCISE 2F

1 The 7.05 train from Glasgow Central to London Euston should take 5 hours 17 minutes for the journey. A points failure at Crewe delays the train for three-quarters of an hour. When does it arrive at Euston?

2 Sonya and Mark play a round of 18 holes of golf in 3 hours 42 minutes. Find the average time per hole, in minutes and seconds, by dividing the total time by 18.

3 The finalists in the 400 metres freestyle race recorded these times:

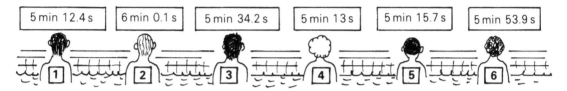

| 5 min 12.4 s | 6 min 0.1 s | 5 min 34.2 s | 5 min 13 s | 5 min 15.7 s | 5 min 53.9 s |

a Calculate the difference in time between swimmers:
 (i) **3** and **5**   (ii) **2** and **6**   (iii) who were the fastest and slowest.
b How much was the winner's time outside the record time of 4 min 57.8 s?

4 A car rally began at 09 45 hours. The winning team took 5 hours 28 minutes to complete the course. At what time did the victors finish?

5 For how long are the sun and the moon above the horizon?

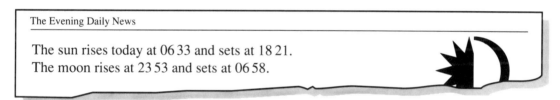

The Evening Daily News

The sun rises today at 06 33 and sets at 18 21.
The moon rises at 23 53 and sets at 06 58.

6 The winner's average lap-time in a *Grand Prix* was 3 min 11.55 s. Calculate the total time for 60 laps in hours, minutes and seconds.

7 The residents of Miller Road were without electricity from 12 48 on April 29th until 10 53 on May 2nd. How long was this?

8 *A puzzle.*   'Four seconds before twenty five to one on the seventh of August 1990'. Can you use each digit from 0 to 9 once to record this time and date?

9 Find out today's 'lights-on' and 'lights-off' times for vehicles. Calculate the number of minutes between:
a lights-off and lights-on   **b** lights-on and lights-off times.

EXERCISE 3F

Extracts from the timetables for two bus routes are shown below.

MELROSE	...	...	09 50	...	10 50	11 50
Tweedbank	...	...	09 57	...	10 57	11 57
Galashiels a	...	...	10 05	...	11 05	12 05
Galashiels d	08 00	09 12	↳	10 12	11 12	12 12
Innerleithen	08 34	09 46	...	10 46	11 46	12 46
PEEBLES a	08 48	10 00	...	11 00	12 00	13 00
PEEBLES d	08 51	10 05	...	11 05	12 05	13 05
Penicuik	09 15	10 31	...	11 31	12 31	13 31
EDINBURGH	19 51	11 07	...	12 07	13 07	14 07

a = arrive
d = depart
↳ = connection

No. 62: MELROSE–
PEEBLES–
EDINBURGH

HAWICK	08 10	09 30	10 30	11 30	12 30
Selkirk a	08 32	09 52	10 52	11 52	12 52
Selkirk d	08 35	09 55	10 55	11 55	12 55
Galashiels a	08 50	10 10	11 10	12 10	13 10
Galashiels d	09 00	10 15	11 15	12 15	13 15
STOW	09 18	10 33	11 33	12 33	13 33
EDINBURGH	10 25	11 40	12 40	13 40	14 40

No. 95: HAWICK–
STOW–
EDINBURGH

1 Sometimes you have to change buses to
 go from one place to another.
 a Which buses would you take for a journey from:
   (i) Tweedbank to Stow
   (ii) Hawick to Innerleithen?
 b Robert takes the 'ten-to-ten' bus at
 Melrose to travel to Penicuik.
   (i) Where does he change buses?
   (ii) How much time has he to make the
     change?
   (iii) What is the earliest time that he can
     reach Penicuik?

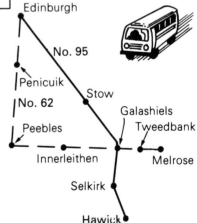

2 Jess lives in Selkirk. She travels to Peebles where her friend can meet her bus at noon.
 a What is the latest bus Jess can catch at Selkirk?
 b Describe her journey, including changes of bus and times involved.

3 Sandy lives in Tweedbank and has to be at a meeting in Edinburgh at 1 pm, at an office which is five minutes walk from the bus station.

   a Give full details of his journey, assuming that he leaves home as late as possible.

   b If he goes straight from the bus to the office, how much time has he to spare before the meeting?

4 It's just past eleven o'clock. Describe two ways of travelling by bus from Melrose to Edinburgh so as to arrive by 2.15 pm. Describe the pro's and con's of each way.

EXERCISE 4F

1 Temperatures are usually measured in degrees Celsius (°C) or degrees Fahrenheit (°F). Scientists sometimes use degrees Kelvin (°K). The flowcharts show how to convert temperatures from one scale to another.

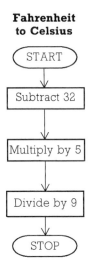

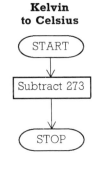

a Construct flowcharts to convert:

   (i) °F to °K

   (ii) °C to °F

   (iii) °K to °F

b Copy and complete:

°F	°C	°K
572	300	573
		773
1202		
	1000	

2 Diane and Jason are using different types of flame to heat the same amount of water.

   a What is the rise in temperature per minute for each person's experiment?

   b Jason claims 'My rise in temperature is greater, so my flame is better'. Why is his claim not justified?

Diane

Time (min)	0	1	2	3	4	5
Temperature (°C)	20	36	52	68	84	100

Jason

Time (min)	0	1	2	3	4	5
Temperature (°F)	68	95	122	149	176	203

   c Convert Jason's readings to degrees Celsius and compare his temperatures with Diane's.

   d What is the room temperature?

EXERCISE 5F

1 Write down the highest and lowest temperatures in each part **a–d**:
  a 0°, 1°, 2°   **b** 0°, 5° − 5°   **c** 10°, 9°, 11°   **d** −6°, −3°, −9°.

2 Calculate the number of degrees between the maximum and minimum temperatures:

a 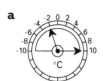   **b**   **c**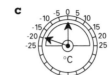

3

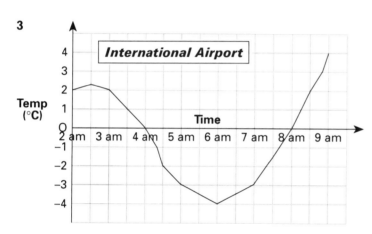

The graph shows the temperature at the airport over a 7 hour period.
  a What was the temperature at 5 am?
  b At what times was the temperature:
      (i) 4°   (ii) −3°?
  c (i) What was the lowest temperature?
      (ii) When was this?
  d For how long was the temperature below zero?

e Copy and complete the table.

Time	2 am	3 am
Temperature °C	2	2

f Use the table to calculate the average hourly temperature from 2 am until 9 am.

4 *Estimate* the temperatures shown on these thermometers.

a    **b**    **c**    **d**    **e**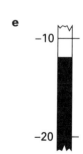

5 Use the flowchart on page 33 to change these temperatures to °C, to 1 decimal place.
  a −40°F   **b** 0°F

7 COORDINATES: X MARKS THE SPOT

EXERCISE 1F

1 Air traffic control must know the positions of aircraft at all times, so it makes use of radar and a coordinate grid. The control tower is at the origin.

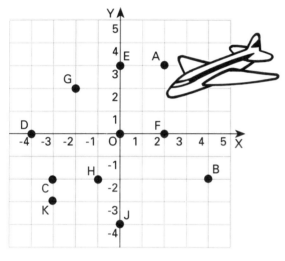

a Which plane is the same distance from the tower as the Airbus at A (2,3)?

b The Boeing 727 at B (..., ...) is twice the distance from the tower as .... What goes in the blank spaces?

c Which plane has the same *x*-coordinate as the Cessna at C?

d The DC-9 is due West of the tower. State its coordinates.

e Three other planes lie North, South, East or West of the tower; the Eagle, the Foxbat and the Jaguar. Identify each and give its coordinates.

f Which two planes are just as far apart as the Gulfstream and the tower?

2 After some time these are the positions of the aircraft.

a Write down the coordinates of each plane.

b List the planes which are:
   (i) closer to the tower than before;
   (ii) further away;
   (iii) still at the same distance.

c Which plane is travelling: (i) fastest (ii) slowest? Explain your answers.

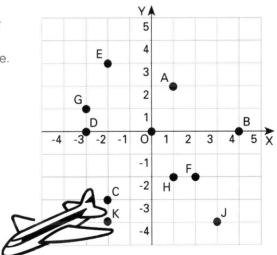

EXERCISE 2F

1 Robo the Robot is 'looking' in a mirror. He can compute his image by changing the signs of the *x*-coordinates of all his points.

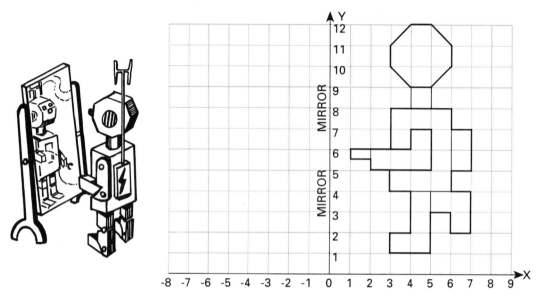

a The points on his head are $(4, 9)$, $(5, 9)$, $(6, 10)$, $(6, 11)$, $(5, 12)$, $(4, 12)$, $(3, 11)$, $(3, 10)$. So its image is at $(-4, 9)$, $(-5, 9)$, $(-6, 10)$, ... Copy and complete this list, and draw Robo and his image on your squared paper.

b List the points on your drawing of Robo's arm and on its image.

c Robo holds up this object to the mirror: $(3, 4)$, $(4, 4)$, $(4, 9)$, $(1, 9)$, $(1, 8)$, $(3, 8)$, $(3, 7)$, $(2, 7)$, $(2, 6)$, $(3, 6)$, $(3, 4)$. Draw the object and its image.

2 What is **t**he **h**andiest **s**ubject? The letters **ths** have been designed on a coordinate grid.

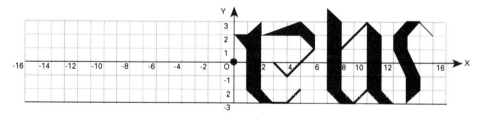

Instructions can be given for drawing each letter. For example, '**t**': $(5, -2) \to$ $(4, -3) \to (3, -3) \to (2, -3) \to (1, -2) \to (1, 1) \to (0, 2) \to (1, 2) \to (2, 3) \to (2, -2) \to (3, -3)$: then $(3, 0) \to (4, -1) \to (6, 1) \to (6, 2) \to (5, 3) \to (2, 2) \to (5, 2) \to (6, 1)$.

a In the same way, give instructions for drawing '**h**' and also '**s**'.

b Copy the letters onto a coordinate grid.

c By changing the sign of every coordinate, new instructions are obtained. For example, changing '**t**' we get $(-5, 2) \to (-4, 3) \to (-3, 3) \to (-2, 3) \to ....$ Follow the new instructions on the same grid to discover **t**he **h**andiest **s**ubject.

EXERCISE 3F

1

A 'little white bull' tiling is being used for a wallpaper pattern. There are pairs of eyes at $(3, 1)$ and $(7, 1)$.

a Write down the coordinates of the next three pairs of eyes in the same line.

b If (x, y) gives a pair of eyes on the line, what can you say about: (i) x (ii) y?

2

Keeping within the range of the diagram in question **1**, give the coordinates of points on the line:

a $y = 3$, where feature Ⓐ occurs;

b $x = 5$, where feature Ⓑ occurs;

c $y = 5$, where feature Ⓒ occurs.

3

A crack in the wall starts at $(4, 0)$, runs along $x = 4$ to $(4, 4)$, follows $y = 4$ to $(8, 4)$, then $x = 8$ to $(8, 6)$, $y = 6$ to $(2, 6)$ and finally $x = 2$ to the top of the wall at $(2, 8)$. Describe the other crack in the same way.

4 Find two points on the lines whose equations are given below. Plot the points, and draw the lines.

a $y = x$ **b** $y = 3x$ **c** $y = 5x$ **d** $y = x + 2$ **e** $y = x - 2$ **f** $y = 3x + 2$

8 SOLVING EQUATIONS

EXERCISE 1F

1 Solve these equations:

a $x - 5 = 11$	**b** $5 - x = 0$	**c** $2y + 1 = 21$	**d** $12 - 2y = 0$
e $7x + 8 = 36$	**f** $5y - 3 = 12$	**g** $11 - 2c = 11$	**h** $64 = 9d + 1$
i $15x + 8 = 68$	**j** $17y + 56 = 56$	**k** $17 + 9t = 107$	**l** $1012 - 23m = 0$
m $\dfrac{10}{k} = 2$	**n** $\dfrac{15}{d} = 5$	**o** $\dfrac{12}{e} = 1$	**p** $\dfrac{24}{t} = 3$
q $11z = 11$	**r** $4m = 2$	**s** $3n = 1$	**t** $6a = 9$

2 Cara is experimenting with her new 'Graphics' calculator. Here are some of her experiments:

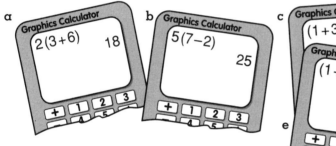 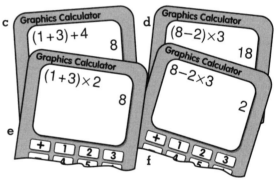

Can you show how she got each answer?

3 Now find the numbers that these letters stand for.

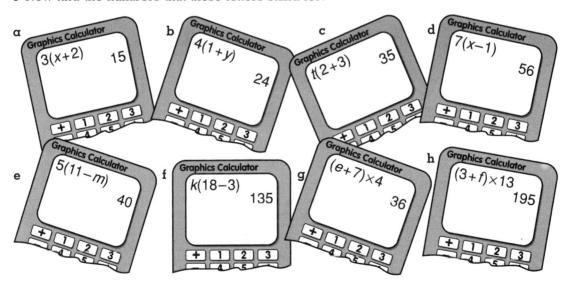

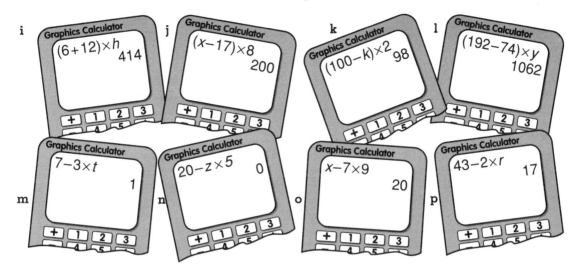

4 Solve this sequence of equations. The solution to each part should help with the next part.

a $\dfrac{420}{p} = 6$ **b** $\dfrac{420}{5q} = 6$ **c** $\dfrac{420}{5(r + 10)} = 6$ **d** $\dfrac{420}{5(s \times s + 10)} = 6$

EXERCISE 2F

---

1 Simplify first, then solve:

a $3x + 2x = 20$ **b** $x + x + x + x = 12$ **c** $7x + 5x = 48$
d $4x + 4x + 1 = 25$ **e** $32 + 5y + y = 38$ **f** $6t + 6 + t = 27$
g $2p + 4p + 6p = 0$ **h** $2k + 1 + k = 16$ **i** $4m - m + 2 = 11$
j $3u - 6 + 2u = 4$ **k** $1 + 4v - v = 7$ **l** $8d + 2 - d + 3 = 19$
m $15 = 2v + 7 - v$ **n** $46 = 3f - 2 + 5f$ **o** $54 = 17 + 3n + 2 + 2n$
p $t + 4t + 2t - 3 = 11$ **q** $3x + 2x + 2 = 17$ **r** $6y + 2y + 1 = 9$
s $2 + t + 3t = 6$ **t** $4n - 3n + 6n = 49$ **u** $5k - k + 2k = 18.$

2 For each situation make an equation, solve it and then say how many litres were in each beaker.

a
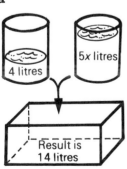

b
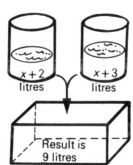

c
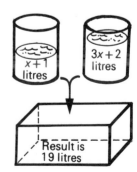

3 In an electrical circuit like this: 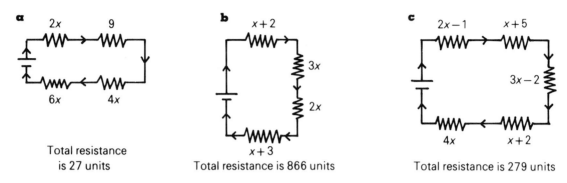 the total resistance is 7 units, the sum of each resistance. 3 units · 4 units

Make an equation for each circuit, and solve it to find the size of each resistance:

a 2x · 9 · 6x · 4x

Total resistance is 27 units

b x+2 · 3x · 2x · x+3

Total resistance is 866 units

c 2x−1 · x+5 · 3x−2 · 4x · x+2

Total resistance is 279 units

4 A jug holds 270 millilitres of juice. The jug fills four bottles, each containing x millilitres; there are 30 millilitres over. Make an equation and solve it to find how much each bottle holds.

5 All the cola in the jug is used to fill bottles A and B: both A and B are then emptied into small bottles, each of which holds x millilitres.

  a Make an equation and find the volume of each small bottle.

  b Calculate the volumes of bottles A and B.

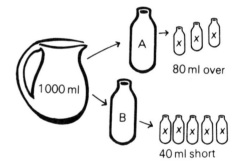

1000 ml

A → 80 ml over

B → 40 ml short

EXERCISE 1F

1 a Write down your estimates for the length and width of your desk, or table, in centimetres.

b Measure the length and width, to the nearest 0.5 cm.

c Calculate the perimeters for **a** and **b**, also the difference between them.

2 Repeat question **1** for the length and width of the classroom, rounding your measurements in **b** to the nearest 10 cm.

3 Repeat question **1** for a large rectangular building, rounding measurements to 0.5 m.

4 Make an accurate drawing of this diagram.

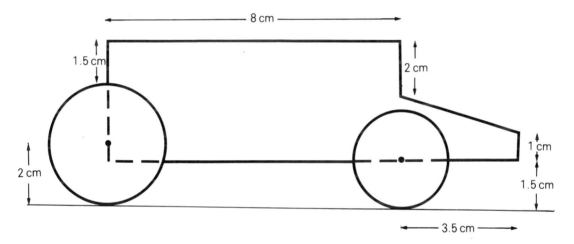

5 1 inch $\doteqdot 2\frac{1}{2}$ cm, 1 yard $\doteqdot$ 1 metre, 1 mile $\doteqdot 1\frac{1}{2}$ km. To 3 decimal places, 1 inch = 2.540 cm, 1 yard = 0.914 m, 1 mile = 1.609 km. Calculate the difference between the approximations for lengths of 100 inches, 100 yards and 100 miles.

EXERCISE 2F

1 Two aircraft, A and B, are flying towards each other at the same height and speed. A is flying east and B is flying west.

A has a radio range of 50 km, and B has a radio range of 70 km.

This drawing shows the position when the aircraft are 140 km apart.

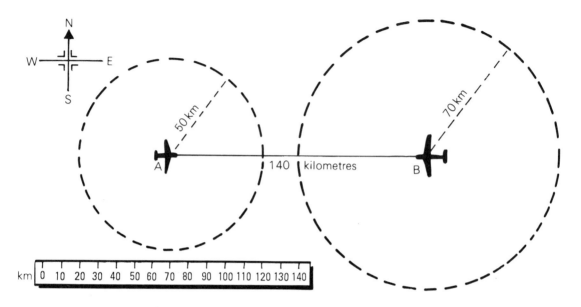

The circles show the limit of radio range of each aircraft.

a Can either pilot hear the other?

b Is it possible for anyone to hear both pilots?

c What is the distance between A and B after the aircraft have flown a further 10 km? Make a scale drawing of the position then, like the one above.

d Mark the position of a person who can now pass messages from one pilot to the other and back again.

e Draw the position after the aircraft have flown a further 10 km.

f Show, by shading, all possible places where radio calls from both pilots can be heard.

g Mark the position of the person furthest north who can receive calls from both pilots. Call this position C.

h How far is: (i) C from A   (ii) C from B   (iii) A from B?

2 A rectangular sheet of wood is cut to measure 3.85 m by 1.40 m.
If the maximum error in both the length and the width is 1 cm, calculate the greatest and least possible perimeters of the wood.

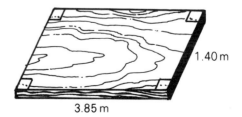

1.40 m

3.85 m

EXERCISE 3F

1 Wood comes from the mill in standard lengths:
Softwood—the range begins at 1.8 m, and increases in steps of 300 mm to 6.3 m.
Hardwood—the range begins at 1.8 m, and increases in steps of 100 mm.
a What are the first three lengths in each range?
b Find the last three lengths in the softwood range.
c How many standard lengths are there in the softwood range?
d The 6.3 m length can be cut into three pieces, each of which has a different standard length. What are the three lengths?

Wallpaper is sold in rolls 10 m long and 533 mm wide. In order to show customers the number of rolls needed for their rooms, Wonderwalls have this table. (Doors and windows are counted as walls.)

Height of room (m)	Perimeter of room (m)					
	10	12	15	17	18	22
2.1–2.3	4	5	6	7	8	9
2.4–2.6	5	6	7	8	9	10
2.7–2.9	5	6	7	9	10	11
3.0–3.2	5	7	8	10	11	12
3.3–3.5	6	8	9	10	12	13
Ceiling	2	2	3	4	5	7

2 Jim's room is 2.8 m high, 5 m long and 4 m wide. How many rolls does he need for the walls?

3 A small room is 3 m long. From the table, 4 rolls are needed for the walls.
a What is the breadth of the room?   **b** How many rolls are needed for the ceiling?

4 Marjorie has a new flat, and decides to paper all the walls and ceilings. The rooms are 2.7 m high. Calculate:
a the number of rolls of wallpaper and ceiling paper for each room (for a perimeter of 14 m, take 15 m in the table, and so on).
b the total cost of the paper.

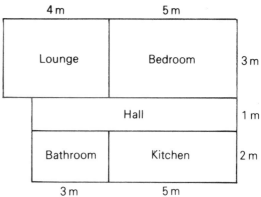

Room	Cost per roll
Lounge	£6
Bedroom	£4
Hall	£3
Kitchen	£3
Bathroom	£2
Ceilings	£2

10 TILING AND SYMMETRY

EXERCISE 1F

1 Copy these floors on squared paper. Tile each floor with the given tile.

a
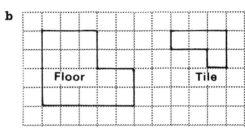

b

2 Only one of these tiles will tile the floor. Which one? On squared paper show how this can be done.

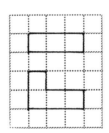 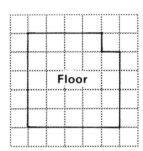

3 Make a tiling pattern on squared paper with each of these tiles.

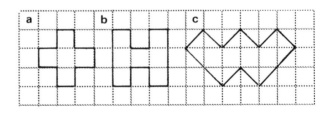

4a For this sequence of tiles, write down the coordinates of the top right-hand corner of the 1st, 2nd, 3rd, 4th, 5th, 6th, 10th and *n*th tiles.
  b Repeat **a** for the bottom left-hand corners of the tiles.

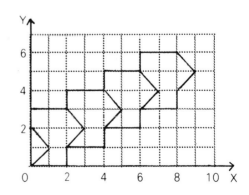

5 Make tiling patterns from these tiles.

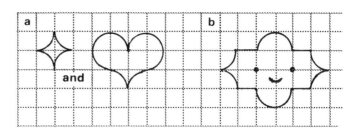

EXERCISE 2F

---

1 How many lines of symmetry has each diagram?

a   **b**   **c**   **d**

e   **f** 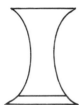   **g**   **h**

2 Copy these diagrams on squared paper, showing the dots and their reflections in the dotted lines.

a 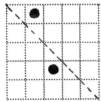   **b** 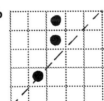   **c** 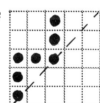   **d**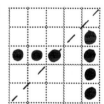

3 These lines of reflection are at right angles. Copy the diagram on squared paper and reflect the dots in *both* lines.

a 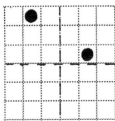   **b** 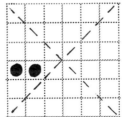   **c**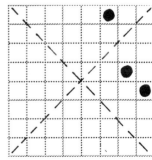

4a Copy and complete the table for the images of the given points when they are reflected in the dotted line.

Point	Image
(1, 3)	
(2, 1)	
(3, 6)	
(4, 0)	
(5, 4)	
(5, 2)	
(a, b)	

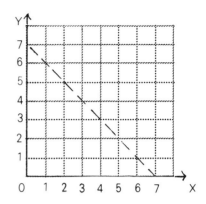

b Copy and complete the table for reflection in the dotted line.

c What is the image of the point (a, b) in the line parallel to the x-axis, through the point:

(i) $(0, 4)$ (ii) $(0, 5)$

(iii) $(0, n)$?

Point	Image
(0, 6)	
(3, 2)	
(4, 1)	
(6, 4)	
(7, 5)	
(1, 3)	
(a, b)	

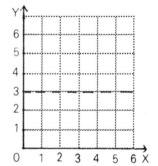

d Investigate the image of (a, b) under reflection in the line through the point $(n, 0)$, parallel to the y-axis.

5 Copy and complete these diagrams so that the dotted lines are axes of symmetry.

a
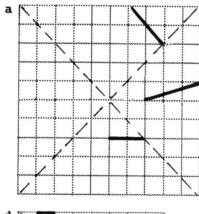

b
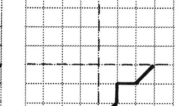

c
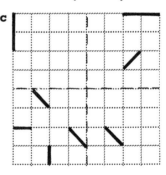

d
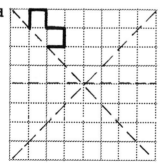

e
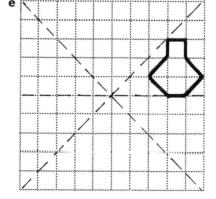

EXERCISE 3F

1 Say whether each diagram has quarter-turn, or only half-turn symmetry, or no turn symmetry at all.

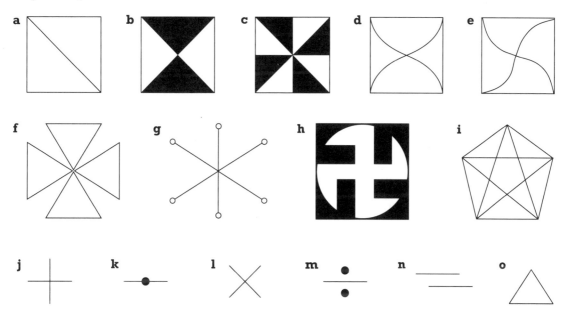

2 Write down the order of symmetry of each shape in question **1**.

3 Copy diagrams **j–o** in question **1**, and mark or find the centre of symmetry of each one.

4 Make shapes from these tiles that have: (i) quarter-turn symmetry   (ii) only half-turn symmetry. Mark the centres of symmetry and any lines of symmetry.

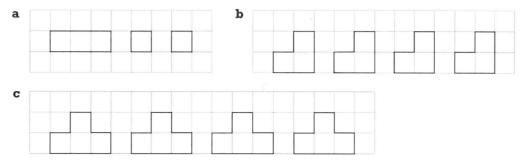

5 Combine these tiles so that the shape formed has:
   a half-turn and line symmetry
   b half-turn but not line symmetry
   c line symmetry but not half-turn symmetry.

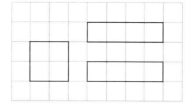

11 MEASURING AREA

EXERCISE 1F

Calculate the shaded areas by adding, or subtracting, the areas of two (or more) rectangles. The lengths are in centimetres, unless otherwise stated.

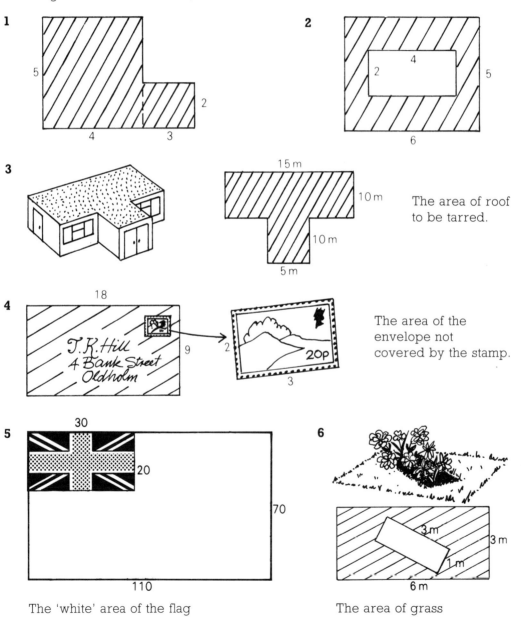

1

5
4   3   2

2

4
2   5
6

3

15 m
10 m
10 m
5 m

The area of roof to be tarred.

4

18
J.K. Hill
4 Bank Street
Oldholm
9   2
20p
3

The area of the envelope not covered by the stamp.

5

30
20
70
110

The 'white' area of the flag

6

3 m
3 m
1 m
6 m

The area of grass

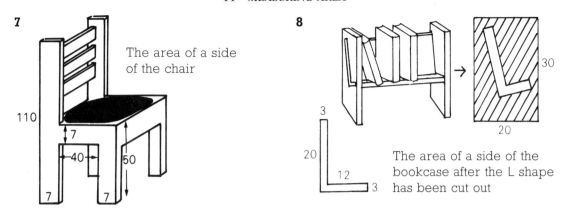

7 The area of a side of the chair

8 The area of a side of the bookcase after the L shape has been cut out

EXERCISE 2F

1 The cost of carpeting a room depends on the area of the floor and the quality of the carpet. Graham and Gilly Gower are planning to put carpet on all the floors in their new flat.

Quality	Cost per m²
Heavy	£25
Medium	£17
Light	£9
Anti-splash	£12

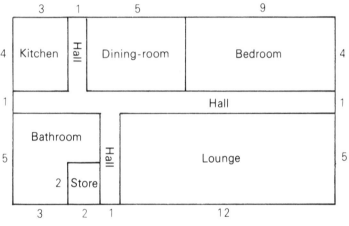

Copy and complete this table to find the total cost.

Room	Area	Quality	Cost per m²	Cost (£)
Bedroom		Light		
Dining room		Medium		
Hall		Heavy		
Kitchen		Anti-splash		
Bathroom		Anti-splash		
Store		Light		
Lounge		Heavy		
			TOTAL	

2 The outside edge of a picture frame is 30 cm long and 23 cm broad.
The width of the frame all round the picture is 2.5 cm.
Calculate:
a the area of the frame
b the cost of the glass to cover the picture, at £35 a square metre.

EXERCISE 3F

1 Calculate the areas of these triangles.

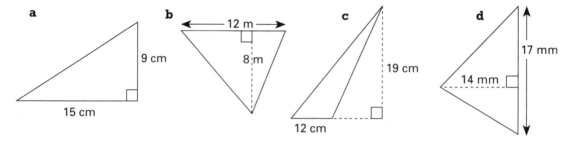

2 Calculate x for each of these sheep pens.

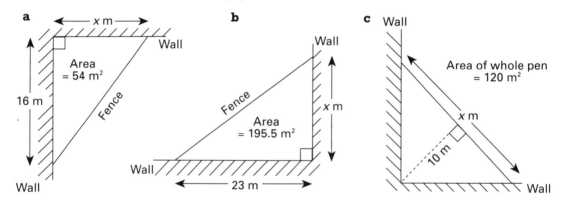

3 Find the area of each of these triangles, using 1 square as a unit:

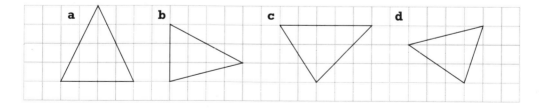

4a Triangle ABC has area 46.41 m².
Calculate the length of AC.

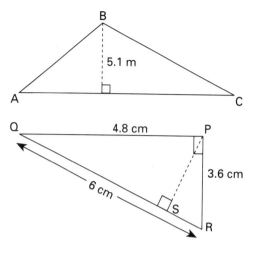

b Calculate the area of triangle PQR, and
then find the length of PS using this area.

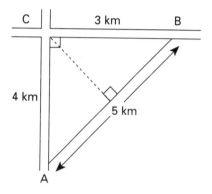

5a Calculate the shortest distance from C to
AB.

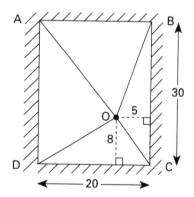

b A rectangular enclosure is called ABCD.
Calculate the area of the part ABOD. The
units are metres.

12 LETTERS, NUMBERS AND SEQUENCES

EXERCISE 1F

1 Find the missing numbers in these tables:

a

a	2	4		6
b	3	5	2	
$2a + b$			8	17

b

k	2	9	3		
$20 - 2k$				8	0

c

x	4	10		20
y	6	20	8	
$\frac{1}{2}(x + y)$			9	30

2 Items in a store are bought for £x each and sold for £y each. Copy and complete the table.

x	70	55		30	$8a$	$11b$
y	100	65	80		$9a$	
Profit (£)			20	5		$3b$

3 Complete the table for the rectangle.

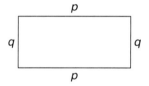

p	6	2	9	5	9
q	4	1	12		7
Perimeter				18	32

4 (i) Make a formula for the perimeter P of each shape.
(ii) Use the formula to calculate the perimeter for the given lengths (in cm).

a

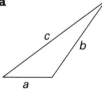

$a = 9$, $b = 13$, $c = 19$

b

$s = 12$, $t = 10$, $u = 14$

c

$k = 12.5$

d

$p = 120$, $r = 80$

5

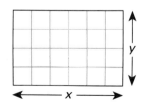

a Find a formula for the length L of wire in this plant support frame, in cm.
b Calculate the total length L when $x = 60$ and $y = 40$.

52

6 Copy and complete these tables:

a

		7	9			11
p	4	1	2	7	2
q	2	3	4	0	
r	1	5	6	9	3
	8	15			

b

a	1	4		4	
$b - a$		1	0		0
b	3		6		
ab			16		
$2a + 3b$					
$3a + 2b$					15

EXERCISE 2F

1 Find the missing entries in column A.

a

	A
1	100
2	200
3	300
10	
25	
n	

b

	A
1	11
2	22
3	33
10	
20	
n	

c

	A
1	8
2	16
3	24
10	
30	
n	

d

	A
1	15
2	30
3	45
10	
25	
n	

e

	A
1	25
2	50
3	75
10	
100	
n	

2

    ● ● ●

Copy and complete the table for the solids shown above.

Number of sides in the base	3	4	5	6	10	n
Number of corners on the solid						
Number of edges on the solid						
Number of faces on the solid						

3   Once again, find the missing entries in column A.

a

	A
1	2
2	3
3	4
⋮	⋮
10	
⋮	⋮
20	
⋮	⋮
n	

b

	A
1	11
2	12
3	13
⋮	⋮
10	
⋮	⋮
15	
⋮	⋮
n	

c

	A
1	21
2	22
3	23
⋮	⋮
10	
⋮	⋮
25	
⋮	⋮
n	

d

	A
1	16
2	17
3	18
⋮	⋮
10	
⋮	⋮
35	
⋮	⋮
n	

e

	A
1	30
2	31
3	32
⋮	⋮
10	
⋮	⋮
27	
⋮	⋮
n	

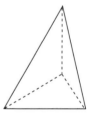

4a Sketch a sequence of pyramids on bases with 3, 4, 5, ... sides.
 b List the number of corners, edges and faces on pyramids with 3, 4, 5, 6, 10, n sides on their bases.

5

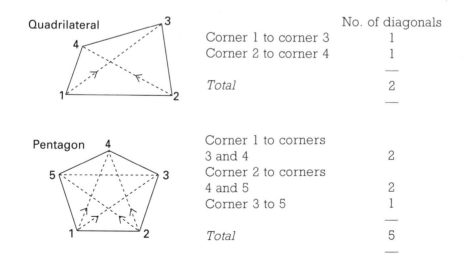

	No. of diagonals
Corner 1 to corner 3	1
Corner 2 to corner 4	1
Total	2

Corner 1 to corners 3 and 4	2
Corner 2 to corners 4 and 5	2
Corner 3 to 5	1
Total	5

 a Draw a hexagon (six sides) and all of its diagonals. Write out the working to calculate the number of diagonals systematically.
 b Write out the working for figures with: (i) 7 sides   (ii) 8 sides   (iii) n sides.

EXERCISE 3F

1 (i) Find the missing entries in column A.
   (ii) If the nth entry in each column is 144, find n each time.

a

	A
1	9
2	18
3	27
10	
n	

b

	A
1	12
2	24
3	36
10	
n	

c

	A
1	18
2	36
3	54
10	
n	

d

	A
1	16
2	17
3	18
10	
n	

e

	A
1	21
2	22
3	23
10	
n	

2 (i) Find the missing entries in column A.
   (ii) For the value of the nth term under each table, find n.

a

	A
1	1
2	4
3	7
10	
n	

b

	A
1	1
2	5
3	9
10	
n	

c

	A
1	5
2	7
3	9
10	
n	

d

	A
1	8
2	14
3	20
10	
n	

e

	A
1	12
2	17
3	22
10	
n	

nth term = 40 nth term = 61 nth term = 81 nth term = 104 nth term = 77

3

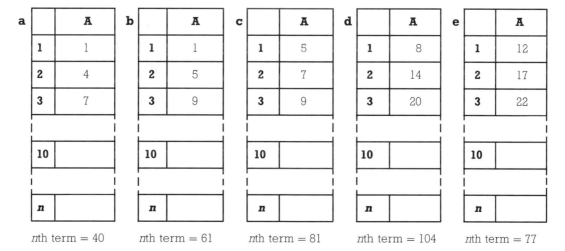

Copy and complete this table for the pattern of hexagons. Each side is 1 cm long.

Number of hexagons	1	2	3	4	10	n
Perimeter in cm						

55

4 Repeat question **3** for a sequence of pentagons.

5 Molly works in the design department of Wonderwalls. She is designing a wallpaper border of triangles with sides of 4 cm.

Make a table of perimeters of patterns with 1, 2, 3, 4, 10 and n triangles.

6a Show that the perimeter of the figure with n 7-sided shapes like these is $5n + 2$ cm.
   b Calculate the perimeter of a strip with 100 shapes in it.
   c How many shapes are needed for a figure with a perimeter of 87 cm?

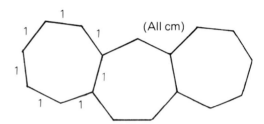

(All cm)

7 Sketch a figure like the one in question **6**, but made of 8-sided shapes.
   a Calculate the perimeter of a strip with 20 shapes in it.
   b How many shapes are there in a figure with a perimeter of 518 cm?

13 TWO DIMENSIONS: RECTANGLE AND SQUARE

EXERCISE 1F

1a Trace these cupboards and drawers.

b Mark, or describe, sets of parallel lines and congruent rectangles.

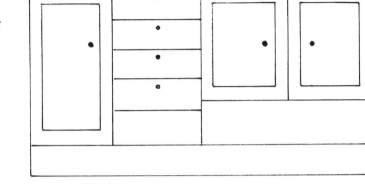

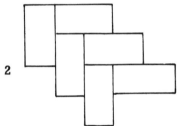

2

a Copy and continue this tiling on plain or squared paper.

b Use arrows to mark sets of parallel lines.

3a Draw each object as you would see it if you looked at it directly from the front.

b Mark sets of parallel lines, and label sets of congruent rectangles with letters.

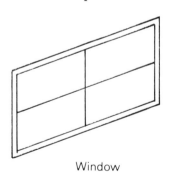

Window

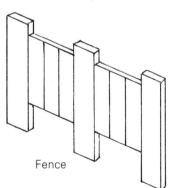

Fence

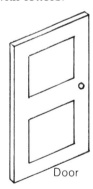

Door

4

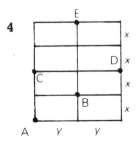

Each road in this rectangular network is x km or y km long. Find the shortest distance from:

a A to E   **b** A to E via B and C   **c** A to E via B, then C and D.

5 The dimensions of the tennis court are shown in feet. Calculate the lengths of AB, BC, DE and EF.

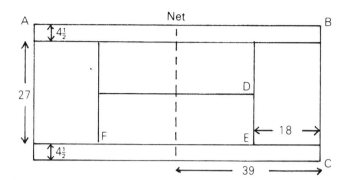

6 PQRS is a rectangle. P is the point (5, 4); PQ is parallel to the *x*-axis and is 4 squares long; PR is parallel to the *y*-axis and is 3 squares long. How many rectangles can you draw on squared paper, using these instructions? Write down the coordinates of the vertices of each one.

7 Four rectangular cards are painted red on one side and white on the other side. How many different patterns can you make when the cards form a tiling like the one shown?

8a In each pattern of tilings, calculate the number of:
    (i) 2 × 1 rectangles
    (ii) 4 × 2 rectangles
    (iii) 6 × 3 rectangles.
b How many 2 × 1, 4 × 2 and 6 × 3 rectangles are there in: (i) the fourth pattern (ii) the *n*th pattern?

EXERCISE 2F

1

The diagram shows two overlapping rectangles which are perpendicular to each other.
a How many right angles can you see?
b How many rectangles are there in the diagram?

2 There are three rectangles here. Copy the diagram.
a Mark in all the right angles.
b Fill in the sizes of all of the other angles.

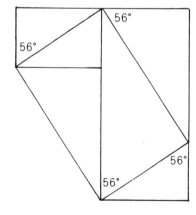

3

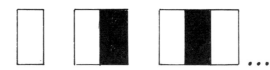

A set of rectangular cards is black (B) on one side, white (W) on the other. A sequence of 1, 2, 3, ... cards is constructed. The first one can be B or W. The second picture can show BB, BW, WB or WW; and so on.

a List the number of possible pictures for:
  (i) 1, 2 and 3 cards   (ii) 4 cards   (iii) 10 cards   (iv) n cards.
b Repeat part **a** if the rectangles are replaced by cuboids with black (B), white (W), red (R) and yellow (Y) sides, standing on square bases.

EXERCISE 3F

1 Stage scenery is supported by rectangular frames of steel tubing. A diagonal tube supports each frame. Calculate the total length of tubing needed here.

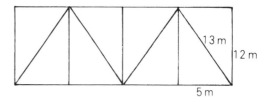

2 EFGH is a rectangle with its sides parallel to the x and y-axes, F is (5, 2) and H (1, 5); FH is 5 units long.
  a What is the length of diagonal EG?
  b Draw the rectangle on squared paper and check your answer to part **a**.

3 Copy the rectangles, and fill in the sizes of as many angles as you can.

a
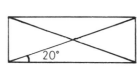

b
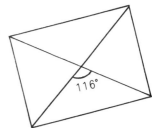

4 The framework consists of two congruent rectangles and AO = 10 cm. Calculate the shortest length along diagonals from
  a A to B   **b** X to Y.

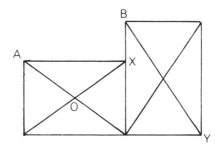

5

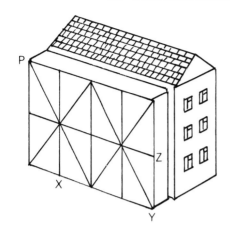

The scaffolding consists of eight congruent rectangles, each 3 m by 4 m, with diagonal supports 5 m long. Calculate:

a the total length of scaffolding

b the shortest distance along the scaffolding from (i) P to X (ii) P to Y (iii) P to Z (iv) P to P, via X then Y then Z.

6

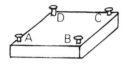

Four terminals A, B, C and D are fixed at the corners of a rectangular board. AB = 24 cm, AD = 18 cm and DB = 30 cm. Each terminal has to be wired directly to every other terminal. What length of wire is needed?

7 A box kite in the shape of a skeleton cuboid is 80 cm long and has a wooden frame with square ends of side 50 cm. Two diagonal struts across each square end are 70 cm long. Sketch the kite, mark the lengths and calculate the total length of wood in the frame.

EXERCISE 4F

1a Write down MOVE and DRAW instructions for drawing a rectangle with its lower left-hand corner at $(0, 2)$, its length parallel to the x-axis and 4 units long and its breadth 6 units long.

b Where do the diagonals cross?

2 A rectangle has its sides parallel to the axes and one diagonal joins $(1, 5)$ with $(6, 2)$.

a What are the coordinates of the other two vertices?

b Write down instructions for drawing the two diagonals.

3 Donna is trying to draw a rectangle, using these instructions. Can you correct the mistake in the instructions?

```
MOVE  (1,2)
DRAW (3,1)
DRAW (5,5)
DRAW (6,3)
DRAW (1,2)
```

4 Give instructions for drawing a rectangle and its diagonals with area 8 cm² and point of intersection of diagonals at $(3, 3)$.

EXERCISE 5F

1 Draw a rectangle PQRS. Mark what you know about its sides and angles. How would you change the rectangle to make it a square?

2 Find the coordinates of N when K (1, 2), L (1, 6), M (5, 6) and N are corners of a square.

3 Find the coordinates of V when S (2, 1), T (5, 3), U (3, 6) and V are corners of a square.

4 Find the coordinates of two possible positions of E and F when C is (4, 7), D is (6, 5) and CDEF is a square.

5 Show four possible positions of the square on one diagram and write down the coordinates of C for each one when A (−2, −1) is one corner of a square ABCD of side 5 units.

6 Draw a rectangle PQRS and its diagonals. Mark what you know about its diagonals. How would you change the diagonals to make PQRS a square?

7 The square sides of the stadium constructed for a pop festival are supported by diagonal struts.
Draw the square and its diagonals, and mark in:
a the lengths of all the lines
b the sizes of all the angles.

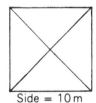

Side = 10 m
Diagonal = 14 m

8 If RSTU is a square, find the coordinates of S and U when R is (3, 1) and T (7, 5).

9 *Sketch* a square on plain paper. How could you check that you have a square? Give as many ways as you can.

10 Show four possible positions of the square and write down the coordinates of C for each one when A (1, −2) is one corner of a square ABCD and the diagonals are 6 units long.

11a Draw a square of side 6 cm. Construct the largest possible circle inside the square. Where is its centre? What is its diameter?
b Draw a circle of radius 8 cm. Construct the largest possible square inside the circle. Explain your method. Measure the length of a side of the square to the nearest mm.

12 A grid of squares is drawn, based on A $(3, 1)$ and B $(8, 1)$.

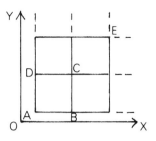

   a Write down the coordinates of the first five vertices in the sequence of corners A, C, E, G, I, ...
   b What are the coordinates of:
   (i) the nth corner
   (ii) the 100th corner, in the sequence A, C, E, G, I, ...?

Investigation

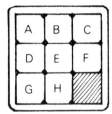

   a Investigate the minimum number of moves needed to take each of these to the bottom right-hand corner:
   (i) E   (ii) G   (iii) D   (iv) A.
   b Repeat part **a** for a 4 by 4 square with letters A–O.

EXERCISE 6F

1 Alana is drawing a nest of squares on a baseline joining $(0, 0)$ and $(8, 0)$. She has completed two squares.

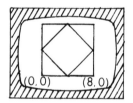

   a On squared paper, or on a screen, draw the two squares shown and four more in the sequence.
   b List instructions for drawing all six squares.

2 Rory wants to draw some shapes on the screen. The computer allows him to visit a point more than once, but not to go along the same path twice or to cross his path at a junction. Write out instructions for drawing these shapes.

a 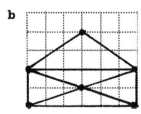   **b**   **c**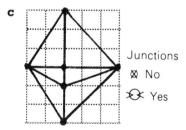

Junctions
⊠ No
⛒ Yes

14 MEASURING VOLUME

EXERCISE 1F

1 Calculate the volumes of these objects by adding, or subtracting, the volumes of two (or more) cuboids. The lengths are in centimetres.

a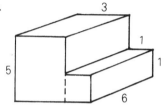

b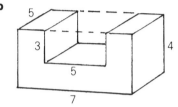

2 How much foam rubber is needed for each of the following? The lengths are in centimetres.

a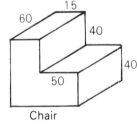

Chair

b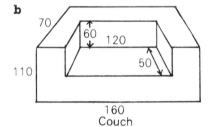

Couch

c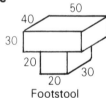

Footstool

3 Calculate the volume of concrete (cm³) needed to build the steps.

4 Calculate the volume of plastic (cm³) for moulding the desk-tidy. (Each hole is 3 cm deep.)

EXERCISE 2F

1 A plastic cube of side 4 cm is melted
down and remoulded as a cuboid.

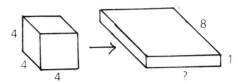

   a Calculate: (i) the volume of the cuboid
                 (ii) the length of the cuboid.
   b Find the dimensions of two other cuboids which can be made from the same cube.

2 A plastic cuboid is melted down and remoulded as the largest possible cube with 'whole
number' sides, along with a smaller cuboid.

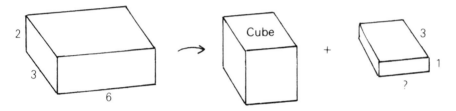

Calculate the length of: **a** the cube   **b** the cuboid.

3 Stewart has a pile of identical cubes. He melts down all but one and remoulds the
material in rectangular sheets 1 cm thick. He uses the sheets to make a box with a lid, so
that the box just holds the cube that he kept. There may, or may not, be some waste
material left over.

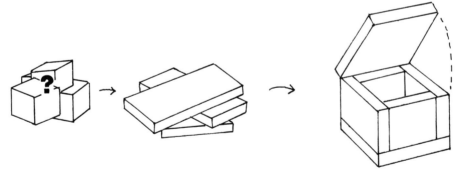

   a How many cubes are needed to make the box if each has a side of
      length: (i) 1 cm (ii) 2 cm (iii) 3 cm (iv) 4 cm?
   b For what sizes of cube will only one be needed to make the box?

15 FRACTIONS AND PERCENTAGES

EXERCISE 1F

1 Where might these fractions be used?
   a $33\frac{1}{3}$ rpm   **b** quarter note   **c** half shut   **d** $2\frac{1}{2}$ turns

2 What fraction of the letters of the alphabet are: **a** vowels   **b** consonants?

3 It is said that shellfish are best to eat when caught in a month which has the letter 'r' in it. What fraction of a year is this?

4 John, Jenny and Jim share a sum of money. Jim has twice as much as Jenny and Jenny has twice as much as John. What fraction of the money does each receive?

5 Explain how a circular piece of paper can be divided into three equal parts by folding only.

6 Which coins are the following fractions of £1?   **a** $\frac{1}{10}$ **b** $\frac{1}{2}$ **c** $\frac{1}{5}$ **d** $\frac{1}{100}$ **e** $\frac{1}{50}$ **f** $\frac{1}{20}$

7 Which unit is: **a** $\frac{1}{360}$ of a complete turn   **b** $\frac{1}{10}$ of a litre   **c** $\frac{1}{1000000}$ of a kilogram?

8a Draw sketches to show how:
   (i) 4 pies could be shared among 5 people
   (ii) 5 pies could be shared among 4 people.
   b What fraction would each person receive in part **a**?

9a How could 3 pieces of chocolate like this be broken so that 4 people would have equal shares?

   b What fraction would each receive?

10 Find two ways in which one triangular piece of chocolate could be broken so that 3 people could have equal shares.

11 Write the following in hours in decimal form. Round your answers to 3 decimal places where necessary.
   a 40 minutes   **b** 48 minutes   **c** 19 minutes   **d** 1 hour 10 minutes
   e 3 hours 13 minutes   **f** 5 hours 59 minutes

12 Find fractions which are equal to: **a** 0.8888...   **b** 0.2727...   **c** 0.1333...

EXERCISE 2F

1 Write each fraction in its simplified form: **a** $\frac{8}{10}$ **b** $\frac{10}{12}$ **c** $\frac{6}{16}$ **d** $\frac{10}{15}$ **e** $\frac{16}{20}$ **f** $\frac{15}{50}$

2 Write each amount as a fraction of £1 in its simplest form:
  a 70p   **b** 25p   **c** 5p   **d** 24p   **e** 95p   **f** 42p

3 The table shows the number of boys and girls in 2B,
and the number with and without glasses.
What fraction of 2B, in simplest form,
  a are boys   **b** are girls
  c wear glasses   **d** are boys who wear glasses
  e are girls who do not wear glasses?

	Boys	Girls	
	6	4	Wear glasses
	8	12	Do not wear glasses

4 What fraction of a circle, in simplest form, is each of these parts?

 a 90°   **b** 120°   **c** 80°   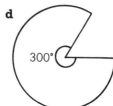 **d** 300°   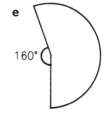 **e** 160°

5 A reminder! ... 1 kilogram = 1000 grams.
Write these weights as fractions of a kilogram in their simplest form:
  a 750 g   **b** 125 g   **c** 1050 g   **d** 2360 g.

EXERCISE 3F

1 A record makes $33\frac{1}{3}$ revolutions per minute. Calculate the number of revolutions in:
  a 3 minutes   **b** 5 minutes.

2 Deepa's tape recorder has three speeds; $7\frac{1}{2}$, $3\frac{3}{4}$ and $1\frac{7}{8}$ inches per second.
  a How many inches of tape will be used at each speed in 8 seconds?
  b What is the connection between the three speeds?

3 Which would you prefer to have: $\frac{5}{6}$ of £30 or $\frac{3}{8}$ of £40?

4 A map has a scale of 1 inch to $3\frac{1}{4}$ miles. Calculate the actual distance between places
which are 4 inches, 10 inches and 36 inches apart on the map.

5 The Dotty Domino factory makes dominoes with lengths between
$2\frac{7}{10}$ and $2\frac{4}{5}$ inches, breadths between $1\frac{4}{5}$ and $1\frac{9}{10}$ inches and
thicknesses between $\frac{1}{5}$ and $\frac{3}{10}$ inch. The dominoes are packed in
boxes in four layers of seven dominoes. Find the inside
dimensions of the smallest and largest boxes needed for a set of
dominoes.

EXERCISE 4F

1 Crumbs and Crackers Bakers break down the cost of making a large wedding cake like this: icing 10%, marzipan 14%, fruit 20%, flour 5%, trimmings 1% and the rest for labour.
  a What percentage of the cost is due to labour?
  b The Company says that 20% of the cost is for fruit and 10% is for icing. This tells us that 'Twice as much is spent on fruit as on icing'. Connect the following in the same way:
  (i) icing and flour   (ii) fruit and flour   (iii) fruit and trimmings   (iv) labour and flour.
  c If the cost of flour is £2, use your statements in part **b** to find the cost of:
  (i) icing   (ii) fruit   (iii) trimmings   (iv) labour.

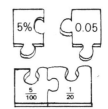

2 These four jigsaw pieces can fit together. They are all equal—the same number expressed as a percentage, as a decimal, as a fraction of 100 and as a fraction in its simplest form. Copy the jigsaws below and complete them in the same way.

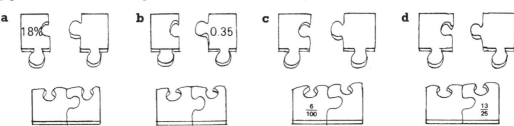

3 Four of these pieces have been matched for you. List as many 'foursomes' as you can and find the missing piece.

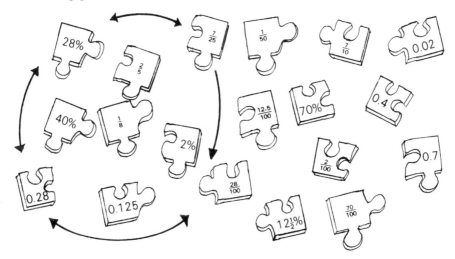

4 In 1990 there were 28 013 000 men and 29 938 000 women in the United Kingdom. Calculate, correct to 1 decimal place, the percentage of:
  a men   **b** women.

EXERCISE 5F

1 This microwave oven costs £249.99, plus
17.5% for Value Added Tax (VAT).
 a Calculate the VAT, and then the total
  cost.
 b Calculate the cost directly by
  multiplying by 1.175.
 c Are your two answers the same? Why?

2 a Which is greater, $\frac{4}{7}$ of £35.28 or $\frac{11}{7}$ of £12.88?   **b** By how much?

3 Josephine has a pay rise of £56 on her monthly salary of £800. Calculate:
 a the fractional increase in her salary   **b** the percentage increase.

4 A small company has five employees. Each month, Adam earns £750, Marjorie £900,
Carol £1040, Dave £580 and Ed £550. Adam's pay is increased by £61.50. Marjorie and
Carol are given the same percentage increase as Adam but Dave and Ed are to have 1%
more. Calculate:
 a Adam's percentage increase   **b** the actual pay increase for each of the others.

5 Mrs Hughes receives $\frac{1}{12}$th of her annual salary of £17 260 for each month she works.
Calculate the amount she receives: **a** in 1 month   **b** in 7 months.

6 Mathematics exam results are graded: 70% and over for A, 60–69% for B, 50–59% for C.
 a Mark scored 19 out of 25 and 26 out of 50 in two tests.
   (i) Change his marks into percentages. What grades did he get?
   (ii) His third test is out of 35. What would he need to score to obtain an 'A' grade
     for his total mark for the three tests?
 b Ann scored 15 in the first test. What would she need to score in the second test in
  order to have a chance of a 'B' grade for her total for all three tests?

7

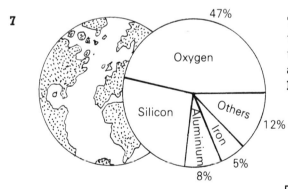

There are elements in the Earth's crust
which can be quantified roughly as shown in
the pie chart.
 a What percentage of the crust is silicon?
 b What weight of silicon, iron and
  aluminium would you expect in a piece of
  crust weighing 10 kg?

 c Calculate the percentage of iron in each
 sample in the table and say whether it is
 above or below the average for the
 crust.

Sample	A	B	C	D
Weight of sample (kg)	1	4	0.5	1.5
Weight of iron (g)	60	184	22	84

16 SOLVING MORE EQUATIONS

EXERCISE 1F

1 Solve these equations:

a $8t + 25 = 49$ **b** $10u + 9 = 9$ **c** $22 - v = 12$ **d** $7v - 5 = 9$

e $17 + 9x = 53$ **f** $79 - 6a = 49$ **g** $17x - 21 = 30$ **h** $23 + 19p = 99$

2 Write in shorter form:

a $x + 3x - x$ **b** $3x - 3x + 4x$ **c** $4x - x + 2x$ **d** $5x - 2x - x$

e $6a + a - a$ **f** $3b - b - b$ **g** $c + 7c - 5c$ **h** $d - d + d$

i $4n + 7n - 5n - 2n$ **j** $6n - 5n + 2n - n + 3n$ **k** $3n - n - n + 2n - n$ **l** $n - n + n - n$

3 Copy and complete **a**, then solve the other equations:

a $t + 4t + 2t - 3 = 11$ **b** $3x + 2x + 2 = 17$ **c** $6y - 2y + 1 = 5$

      $7t - 3 = 11$ **d** $2 + t + 3t = 6$ **e** $4n - 3n + 6n = 49$

          $7t =$ **f** $5k - k + 2k = 18$ **g** $n - n + 6n - 4 = 20$

            $t =$ **h** $7x + 7x - 2x = 60$ **i** $y + 9y - y + 7 = 43$

4 Solve these equations by 'covering up' the contents of the brackets:

a $3(x + 4) = 15$ **b** $2(y - 2) = 6$ **c** $4(z + 1) = 24$ **d** $5(t - 3) = 20$

e $10 - (x + 1) = 7$ **f** $8 + (1 - x) = 8$ **g** $(x + 3) - 7 = 4$ **h** $(2x - 6) + 1 = 11$

EXERCISE 2F

1 Calculate:

a $\begin{array}{r} 5x \\ +6x \\ \hline \\ \hline \end{array}$ **b** $\begin{array}{r} 8x \\ -7x \\ \hline \\ \hline \end{array}$ **c** $\begin{array}{r} 9y \\ +7y \\ \hline \\ \hline \end{array}$ **d** $\begin{array}{r} 10y \\ -4y \\ \hline \\ \hline \end{array}$ **e** $\begin{array}{r} 8z \\ -8z \\ \hline \\ \hline \end{array}$

f $\begin{array}{r} 2x + 1 \\ -2x \\ \hline \\ \hline \end{array}$ **g** $\begin{array}{r} y + 3 \\ -y \\ \hline \\ \hline \end{array}$ **h** $\begin{array}{r} 5t - 1 \\ -t \\ \hline \\ \hline \end{array}$ **i** $\begin{array}{r} 4s + 3 \\ +s \\ \hline \\ \hline \end{array}$ **j** $\begin{array}{r} 5m + 3 \\ -5m \\ \hline \\ \hline \end{array}$

2 Copy each expression, then:

(i) show the action to cancel out the letter-term

(ii) fill in the result.

Example $\begin{array}{r} 3x + 2 \\ -3x \\ \hline 2 \\ \hline \end{array}$

a $2x + 5$ **b** $8x + 1$ **c** $3 - 2y$ **d** $4 - m$ **e** $m + 4$

f $10 - t$ **g** $4k + 4$ **h** $1 - x$ **i** $13a + 9$ **j** $15 - 13k$

3 Simplify:

  a $5 + 3x - 3x$ **b** $2y + 1 - 2y$ **c** $6d + d - 7d$ **d** $5p - p - 4p$ **e** $3t - 1 + 4 - 3t$

4 Write down: (i) the number of weights on each tray
                    (ii) the number after the given action.

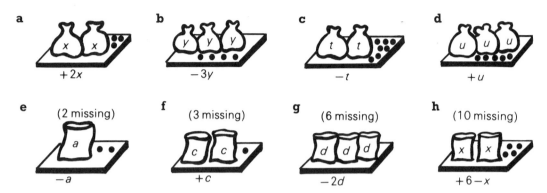

a $+2x$ **b** $-3y$ **c** $-t$ **d** $+u$

e (2 missing) $-a$ **f** (3 missing) $+c$ **g** (6 missing) $-2d$ **h** (10 missing) $+6-x$

EXERCISE 3F

1 Solve these equations:

  a $5x = 2x + 12$ **b** $8y = 9 - y$ **c** $3p = 10 - 2p$ **d** $7k - 4 = 3k$
  e $9t = 2t + 35$ **f** $v = 2 - v$ **g** $3y - 10 = y$ **h** $15u = 72 + 9u$

2 Make an equation for each balance and solve it, to find the weight in each bag, or the weight of an apple, orange, and so on.

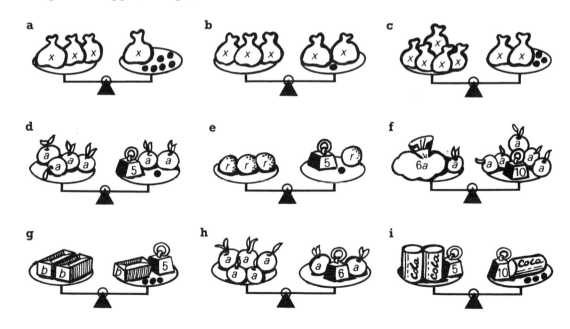

3 In each pair of columns, the numbers in the nth row are equal. Find which row this is, by solving an equation.

a

	A		**B**
1	4		22
2	8		24
3	12		26
4			
5			
n			

b

	A		**B**
1	7		63
2	14		66
3	21		69
4			
5			
n			

c

	A		**B**
1	12		45
2	24		46
3	36		47
4			
5			
n			

4 Solve:
a $6x + 12 = 30$ **b** $13y - 7 = 45$ **c** $7z + 9 = 128$ **d** $5a - 3a + 7a = 81$
e $4b - 3b + 2b - b = 50$ **f** $c + 8c - 6c = 42$ **g** $7 = 13 - 2k$ **h** $14 = 38 - 2m$
i $6t - 50 = 4t$ **j** $8u = u + 91$ **k** $36 - 11v = 7v$ **l** $9u + 1 = 3u + 13$

EXERCISE 4F

1 Solve:
a $5p - 8 = 3p + 6$ **b** $7q - 3 = 2q + 12$ **c** $6r + 1 = r + 36$ **d** $2s + s + 1 = s + 5$
e $4t - t = 2t + 9$ **f** $7u - 2u + 3 = u + 7$ **g** $6 - 2g = 5g - 22$ **h** $2 - 7h = 12 - 12h$

2 Make an equation for each picture, and solve it to find x, a, c, ... The first bag means $x - 3$.

a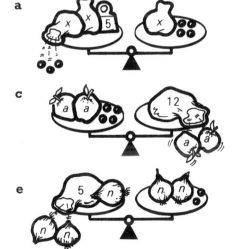

b

c

d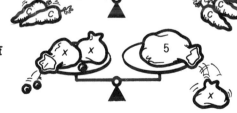

e

f

3 Each pair of pictures shows two ways in which I can spend all my spare money. Make an equation for each pair and solve it to find how much money I have in each case.

a

Onion rings cost 15p per portion.
(Let *h* pence = cost of 1 hamburger.)

b

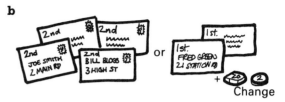

First class costs 5p more than second class.
(Let *s* pence = cost of second class stamp.)

c

An orange costs 2p more than an apple.
(Let *a* pence = cost of 1 apple.)

d

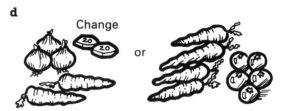

An onion costs 5p more than a carrot.
A tomato costs 2p more than an onion.
(Let *c* pence = cost of 1 carrot.)

17 THREE DIMENSIONS

EXERCISE 1F

1 The box that Jack lives in has many
  properties. Which of the following help
  you to identify it as a cuboid?
  a The box is yellow with red writing.
  b All its faces are rectangles.
  c The word 'Jack' is written on it.
  d The word 'Jack' looks the same when it
  is inverted.
  e It has six faces.
  f It has a lid.
  g It has eight corners.
  h Its volume is 480 cm³.
  i All the angles are right angles.
  j It has twelve edges.
  k The box looks the same when it is
  inverted.

2 Believe it or not, each of these solids shares at least one property about the number of
  faces, corners or edges with a cuboid. Find the properties.

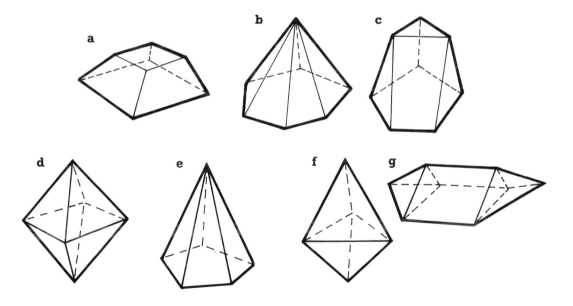

3 List some facts about a cube that are not true for a cuboid.

EXERCISE 2F

Practical

1 Here is a clever trick for drawing solids. It is called perspective drawing. Follow these steps to draw two large blocks of flats.

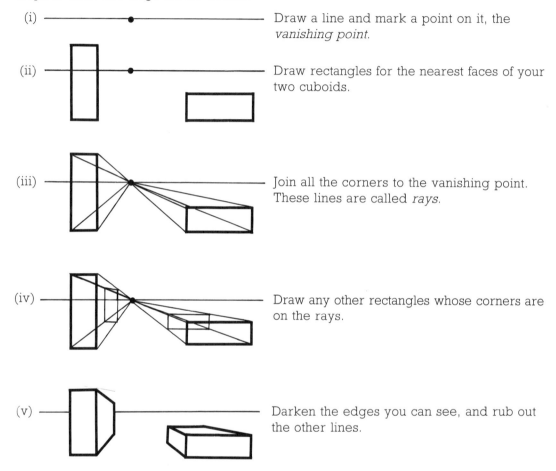

(i) Draw a line and mark a point on it, the *vanishing point*.

(ii) Draw rectangles for the nearest faces of your two cuboids.

(iii) Join all the corners to the vanishing point. These lines are called *rays*.

(iv) Draw any other rectangles whose corners are on the rays.

(v) Darken the edges you can see, and rub out the other lines.

2 Practise the technique on your own, starting rectangles at different places—above, on or below eye level; and to the left, right or on the vanishing point.

3 Construct a street scene or a room of furniture using perspective drawing.

4 Investigate how to use two vanishing points.

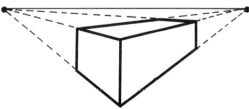

EXERCISE 3, 4F

1 W.E. Boxem make small boxes from rectangular sheets of card 30 cm by 20 cm. Their machines can only cut out corners to make nets for different sizes of box.

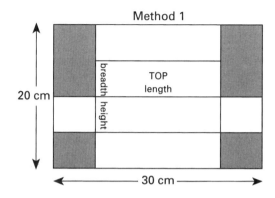

Method 1

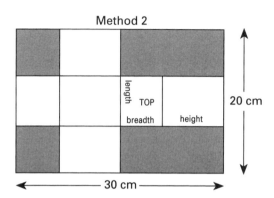

Method 2

a By sketching the net for Method 1 and marking the length, breadth and height, find which of the following order sizes for boxes the firm can accept. The figures give length × breadth × height, in whole numbers of centimetres.
(i) $12 \times 9 \times 1$ (ii) $14 \times 8 \times 2$ (iii) $15 \times 7 \times 3$ (iv) $18 \times 6 \times 4$ (v) $20 \times 5 \times 4$.

b Repeat part **a** for Method 2 and these lengths, breadths and heights (in that order):
(i) $7 \times 6 \times 9$ (ii) $10 \times 5 \times 10$ (iii) $12 \times 4 \times 11$ (iv) $14 \times 3 \times 10$ (v) $16 \times 2 \times 13$.

c If the length is greater than the breadth, Method 1 produces 9 different boxes and Method 2 produces 6.
(i) Can you find their lengths, breadths, heights and surface areas?
(*Hint* For Method 1 you could use $L + 2B = 30$, $B + H = 10$. Be systematic—take $B = 1, 2, 3, \ldots$ make tables of figures.)
(ii) Which dimensions give the lightest box?

2 Suit Yourself Cards plc make special designs for their boxes of cards. This is the net of one of their boxes.

The box is put in a presentation case which shows only the top half of one of the faces.

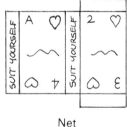

Net

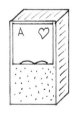

Presentation case

a How many different 'views' of the box are possible as it is packed in different ways?
b In fitting the box into the case, what properties are being used of:
(i) a rectangle   (ii) a cuboid?

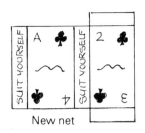

New net

New presentation case

c A new net is made and a box of each type is packed in a new presentation case. How many different views are now possible in:
   (i) one space in the case, using either box
   (ii) the second space, after the first one is filled?

d (i) Explain why 32 views are possible in the type of case in question **c**.
   (ii) List them as (AH, AC), (AH, 2C), (AH, 3C), (AH, 4C) and so on.

e Deluxe cases of three and four boxes with different designs are being prepared. How many of each can be packed before a pattern is repeated?

EXERCISE 5F

A piece of thin polythene tubing is made into an open-ended cuboid (as for a box kite) by slipping it tightly over a skeleton cuboid. Any polythene over is tucked in at one end, as shown.

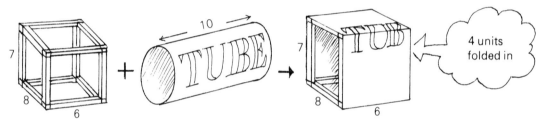

1 What is the perimeter of the opening of the tubing?

2 Which of these cuboids can be fitted tightly with the tubing? What length of tubing has to be tucked in?

a 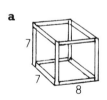   **b** 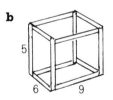   **c** 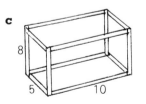   **d**

e 7 × 4 × 8 cuboid   **f** 3 × 4 × 11 cuboid   **g** 9 × 10 × 4 cuboid   **h** 12 × 3 × 10 cuboid

3 a Draw three more skeleton cuboids that would provide suitable frameworks.
   b Give rules for recognising suitable frameworks.

EXERCISE 6F

1 Sketch the edges of this room as if you are looking into a cuboid.

   a Label the nearest 'wall' ABCD, the floor ABFE and the right-hand wall BFGC. The remaining corner is H.

   b Name the face which represents:
     (i) the back wall (ii) the ceiling (iii) the left-hand wall.

   c Describe the position of H.

2 Sara sketched her TV set as a cuboid, but forgot to name all the corners.

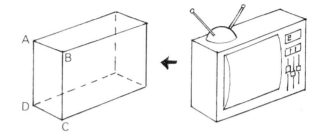

   a If ABCD and PQRS are faces, in how many ways can the cuboid be labelled?

   b If PB is one edge, how many ways now are there to label it?

   c If ABPQ is the top, name all the sets of equal edges, then draw and label the cuboid.

18 PROBABILITY

EXERCISE 1F

Unlikely		Likely	
Impossible ─────────────── Even chance ─────────────── Certain

Choose the best word or words to describe the chance of each of these events taking place.

1 The population of your town or city will increase next year.

2 Obtaining an even number on one roll of a dice.

3 Choosing at random a month of the year which doesn't contain the letter E.

4 A golfer having two holes in one, at successive holes.

5

13	4	21
60	92	35

Choosing a card at random which will give you:
a an even number   **b** an odd number
c a number divisible by 3   **d** a number greater than 10.

6 You will hear the clap of thunder before you see the flash of lightning.

7 Having tossed a coin ten times and obtained ten heads, you will obtain a head on the next toss of the coin.

8 You will be a millionaire before you are thirty.

9 An acid will turn litmus paper red.

10 A person chosen at random will have an above average IQ.

EXERCISE 2F

1 A letter is chosen at random from the word DIVISION. What is the probability that the letter chosen is:
  a S   **b** I   **c** a vowel   **d** a consonant?

2 Rashid posts 8 letters: 2 to France, 3 to the USA, 2 to India and 1 to Australia. One of the letters does not arrive. Calculate the probability that the missing letter was bound for:
  a Australia   **b** the USA   **c** France.

3 a City have played 36 matches this season so far. Estimate the probability that they will win their next game.
  b Estimate the probability that United will:
    (i) win (ii) lose (iii) draw their next match.
  c What assumption did you make in **a** and **b**?

	P	W	L	D
United	36	24	6	6
City	36	12	12	12

4 Tanya has applied to sit her driving test in June. She is unable to sit it on a Tuesday or Friday. The day, excluding Sunday, is chosen at random. Calculate the probability that Tanya is given a date which is unsuitable.

June					
S		6	13	20	27
M		7	14	21	28
T	1	8	15	22	29
W	2	9	16	23	30
T	3	10	17	24	
F	4	11	18	25	
S	5	12	19	26	

5

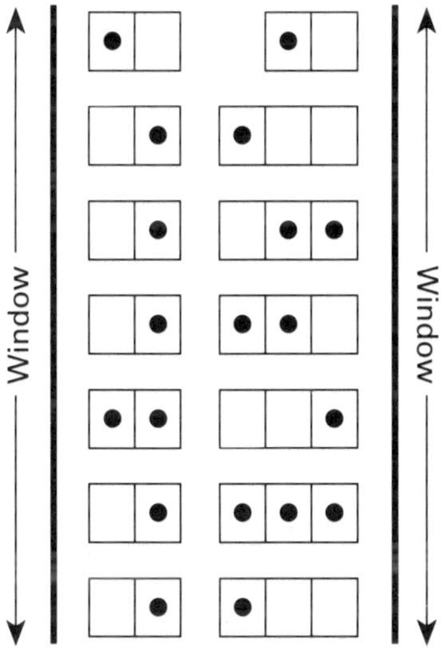

The diagram shows the seats that are occupied in a train compartment. A passenger enters the compartment and sits at random. Calculate the probability that she sits:
  a at the window
  b by herself.

6 A palindromic number is one which reads the same backwards, e.g. 171. A number is selected at random from 10 to 1000 inclusive. What is the probability that it is a palindromic number?

7 a Norman has a problem. He is not sure which package goes into which envelope.
   (i) How many possible outcomes are there?
   (ii) What is the probability that he puts the correct packages into the correct envelopes?

b He now has a bigger problem. What is the probability that he puts the correct packages into the correct envelopes this time?

8 Hannah has a main course and dessert for her lunch. She likes all the items on the menu.
   a How many possibilities are there for her lunch? List them.
   b What is the probability that she chooses curry and apple crumble?

9

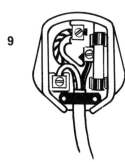

   a How many ways are there of connecting the three wires to the plug?
   b What is the probability of choosing the wrong way of wiring the plug, if no instructions are given?

ANSWERS

1 WHOLE NUMBERS IN ACTION

PAGE 1 EXERCISE 1F

1 72 **2a** Julie 207, Nicholas 288 **b** Julie 6, Nicholas 2
3a 0 **b** 217 **4a** 4 **b** 0 **5** 125 km
6a 44 km **b** (i) 69 km (ii) 56 km
7 25 km **8** 103 km

PAGE 2 EXERCISE 2F

1a 14, 19, 12; 13, 15, 17; 18, 11, 16 **b** 34, 33, 38; 39,
35, 31; 32, 37, 36 **c** 4, 9, 5, 16; 15, 6, 10, 3; 14, 7,
11, 2; 1, 12, 8, 13
3 Missing numbers: **a** 20, 3, 24; 9, 17; 2, 23; 12; 14,
18 **b** 2, 3; 7, 12; 22; 20; 30, 28; 33, 5
4 Missing numbers: 97, 96; 12, 88, 14; 25, 71; 40; 42,
57; 50, 43, 54, 59; 38, 35; 76; 81; 2

PAGE 3 EXERCISE 3F

1 1 million = 1 thousand thousands; 1 hundred
hundreds = ten thousands; ten hundreds = 1 thousand
2a $15 \div 3 \times 7 = 35$ **b** $28 \div 4 \times 3 = 21$
c $27 \div 9 \times 8 = 24$ **d** $36 \div 9 \times 5 = 20$
e $48 \div 16 \times 2 = 6$ **f** $56 \div 8 \times 6 = 42$,
$56 \div 4 \times 3 = 42$
3 $1 + 3 + 6 + 8 = 2 + 4 + 5 + 7$;
$1 + 2 + 7 + 8 = 3 + 4 + 5 + 6$;
$1 + 4 + 5 + 8 = 2 + 3 + 6 + 7$
4a 10 **b** 20
5a $1 \times 2 + 3 + 4 = 9$; $1 \times 2 \times 3 + 4 = 10$;
$1 + 2 \times 3 + 4 = 11$; $1 + 3 + 2 \times 4 = 12$;
$1 \times 2 + 3 \times 4 = 14$; $1 + 2 + 3 \times 4 = 15$;
$1 \times 2 \times 3 \times 4 = 24$; $1 + 2 \times 3 \times 4 = 25$ **b** (i) Extra
numbers: $13 = 4 + 3 \times (1 + 2)$; $16 = 2 \times (1 + 3 + 4)$;
$17 = (1 + 4) \times 3 + 2$; $18 = 4 \times (3 + 1) + 2$;
$19 = (2 + 4) \times 3 + 1$; $20 = 1 \times 4 \times (2 + 3)$;
$21 = (1 + 2 + 4) \times 3$ (ii) $36 = 4 \times 3 \times (2 + 1)$

PAGE 4 EXERCISE 4F

1a 22, 29. Add 7 **b** 35, 44. Add 9 **c** 26, 37. Add
1, 3, 5, ... **d** 66, 60, ... Subtract 10, 9, 8,
... **e** 27, 81. Multiply by 3 **f** 9, 11. Add 1, 2, 1,
2, ...
3a 25, 36, 49; 9, 11, 13; 2, 2, 2 **b** Subtract the two
terms in the row above
4a 6×4, 7×5 dots **b** (i) 4×6, 5×7, 10×12,
100×102 (ii) 4×8, 5×10, 10×20, 100×200
5a 15 **b** 55 **c** 5050

PAGE 5 EXERCISE 5F

1a 500 **b** 0 **c** 3300 **d** 40 000 **e** 1 000 000
f 1 000 000 **g** 10 000 **h** 10 000 000

2a 160 **b** 424 **c** 450 **d** 4040 **e** 0 **f** 400
g 1360 **h** 35 000 **i** 490 000 **j** 2 400 000
k 75 600 **l** 179 200
3a 77 **b** 297 **c** 570 **d** 5050 **e** 0
f 10 000 **g** 66 660 **h** 142 100 **i** 100 000
j 258 000 **k** 1210 **l** 1111'
4a 900 **b** 340 **c** 72 000 **d** 4950 **e** 100 000
f 216 000 **g** 7400 **h** 8910 **i** 64 000
j 720 000 **k** 27 000 000 **l** 0

PAGE 5 EXERCISE 6F

1a (i) 6000 miles (ii) 5000 miles **b** 11 000
miles **c** 10 886 miles **d a** (i) 11 000 miles
(ii) 10 000 miles **b** 21 000 miles **c** 20 357 miles
2a (i) Up 3463 (ii) 614 970 (iii) 1 011 605
(iv) 683 466 **b** (i) 2 708 621 (ii) 2 759 981
3a (i) 8 (ii) 16 **b** (i) 80, 205, 336, 430, 220, 45
(ii) 1316 **4** £4.90, 70p **5a** (i) 40 (ii) 80 **b** 2

2 ANGLES AROUND US

PAGE 7 EXERCISE 1F

1a (i) It is half the angle round the point B, that is $\frac{1}{2}$ of
$360°$, or a straight angle. (ii) 90 is half $\angle$ ABC.
2a Acute: (ii), (iii), (v). Obtuse: (i), (iv), (vi)
b Fold AB along DB to get an angle of $45°$ for angle
(iii). Open out to $180°$, then fold over $45°$ to get $135°$
for angle (iv). **3a** (i) 1 (ii) 3 (iii) 5
4 Acute acute acute,
acute acute right,
acute acute obtuse

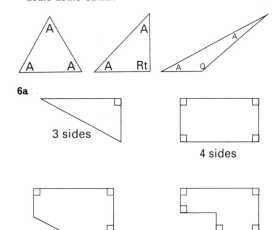

b Sides: 6 7 8 9 10 11 12 13 14
Rt $\angle$s: 5 5 6 7 7 8 9 9 10
Pattern: (5, 5, 6), (7, 7, 8), etc.

PAGE 8 EXERCISE 2F

1a Acute: ∠AOB, ∠BOC; obtuse: ∠AOC, ∠DOC, ∠DOA, ∠BOD   **b** Acute: ∠AOB; obtuse: ∠AOC, ∠AOD   **c** Reflex angles AOB, BOC, COD, AOD, AOC, BOD
2 (i) $a = 40$, $b = 100$ (ii) $a = 30$, $b = 120$, $c = 60$, $d = 60$
3 6   **4a** 9, 6   **b** $1\frac{1}{2}$ **5a** (i) 4   (ii) 8   (iii) 8
b $p = 2q$, $p = 2r$, $q = r$ **c** (i) s, 4   (ii) t, 12
(iii) u, 6; $s = 3t$, $u = 2t$, $s = 1\frac{1}{2}u$.
6 40° or 100°; two answers.
7a (i) 80°   (ii) 60°   (iii) 140°   (iv) 120°
b (i) 103°   (ii) 51°   (iii) 129°   (iv) 154°
8a 135°   **b** 4
9a 360°   **b** 45°   **c** $\frac{1}{2}$ turn   **d** 8 turns

PAGE 10 EXERCISE 3F

1a 6   **b** 3   **2a** (i) 30°   (ii) 70°   **b** 80°
3a (i) 90°   (ii) 80°   (iii) 110°   **b** 80°
4b The dotted lines bisect the angles.
5a AB = AD, CB = CD, ∠ABC = ∠ADC   **b** 180°

PAGE 11 EXERCISE 4F

1b The angles add up to 180°
2 (*Rows down*) v: 140, 90, 55, $u + v$: 180, 180, 180

PAGE 12 EXERCISE 5F

1a Same as for cube.
b 3 horizontal (h), 0 vertical (v); $1h$, $1v$; $0h$, $0v$; $1h$, $0v$
c $4h$, $0v$; $2h$, $2v$; $0h$, $0v$; $0h$, $1v$; $1h$, $1v$ (for some)
2a At noon and midnight   **b** Never
c Between 12–1 twice, 1–2 twice, 2–3 twice, 3 once, 3–4 once, 4–5 and 5–6 and 6–7 and 7–8 twice each, 8–9 once, 9 once, 9–10 once, 10–11 and 11–12 twice each
3 The cube and the cuboid each have three sets of four parallel edges; and each edge has four perpendicular intersecting edges.
The tetrahedron has no parallel or perpendicular edges. The square pyramid has two pairs of parallel edges as the base (a square); and each edge of the square base has two perpendicular edges.
4a Four vertical, twelve horizontal   **b** No vertical, 6 horizontal; or 2 vertical, 6 horizontal; or no vertical, 8 horizontal.

3 LETTERS AND NUMBERS

PAGE 13 EXERCISE 1F

1a $x = 8$ **b** $y = 26$ **c** $t = 14$ **d** $x = 6$
e $y = 7$ **f** $x = 7$ **g** $w = 12$ **h** $t = 6$
i $x = 9$ **j** $x = 12$, $y = 8$
2a $x = 12$ **b** $y = 3$ **c** $x = 1$ **d** $t = 8$

3a $h = 8$, $j = 34$, $g = 43$, $a = 3$, $c = 7$, $d = 19$, $b = 7$, $e = 27$, $f = 31$ **b** $x = 4$, $t = 55$, $a = 88$, $w = 58$, $b = 172$, $y = 49$, $r = 211$, $c = 340$, $e = 226$

PAGE 15 EXERCISE 2F

1a (i) $x + 2$ (ii) $x + 5$ (iii) $x + 1$ (iv) $x + 13$
(v) $x - 1$ (vi) $x + 4$ **b** (i) $y - 3$, $y - 1$, $y + 1$ or $y + 3$ (ii) $y - 1$
c Go 5 houses along on the same side, to the East
2a   **b** $13 - x$

page 12	page 1	page 10	page 3	page 8	page 5

c   **d** $21 - x$

page 14 − x	page $x − 1$		page 22 − x	page $x − 1$

3a (i) II or III; both have rule 'add 1' along the rows.
(ii) I or V; both have rule 'subtract 1' down the columns.
(iii) V   (iv) II   (v) I or II   (vi) II
b (i) VI, $w = 18$ (ii) IV, $y = 18$ (iii) IV, $x = 24$

PAGE 18 EXERCISE 3F

1a $x + 5$, $x + 2$, $x + 9$, x, $x + 6$
b $2x$, $2x + 7$, $3x + 7$, $2x$, $3x + 1$
2a $2t + 2$ **b** $4t + 10$ **c** $2t + 2$ **d** $4t$
e $3t + 2$ **f** $4t + 10$ **g** $3t + 2$ **h** $4t$
a, c; b, f; d, h; e, g
3 *Ann* $6m + 4$, *Mary* $6m + 8$, *Ian* $6m - 4$, *Sue* $6m - 4$, *Ian's* and *Sue's* are the same
4a (i) *Rows:* 1, 3, 5, 8, 9; 4, 12, 14, 26, 27; 6, 18, 20, 38, 39; 9, 27, 29, 56, 57
(ii) *Rows:* 1, 2, 7, 11, 9; 4, 8, 13, 29, 27; 6, 12, 17, 41, 39; 9, 18, 23, 59, 57
b $x \to 3x \to 3x + 2 \to 6x + 2 \to 6x + 3$, for (i)
$x \to 2x \to 2x + 5 \to 6x + 5 \to 6x + 3$, for (ii)
The output cells are equal
5a (i) 18, $3x$, 8   (ii) 12, $5x$, 16, x
b for (i), $7x + 26$; for (ii), $6x + 28$ **c** $x = 2$

4 DECIMALS IN ACTION

PAGE 20 EXERCISE 1, 2F

1a 700   **b** 60   **c** $\frac{4}{10}$ **d** $\frac{2}{1000}$
2 12.10, 12.01, 10.05, 9.97, 9.79
3a 60, 3, $\frac{5}{10}$, $\frac{4}{100}$, $\frac{6}{1000}$ **b** 20, 6, $\frac{9}{10}$, $\frac{80}{100}$, $\frac{1}{1000}$, $\frac{5}{10000}$, $\frac{4}{100000}$
4a 1.6   **b** 2.4   **c** 3.2   **d** 5.5   **e** 7.5
f 2.032   **g** 2.048   **h** 2.052

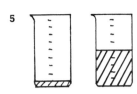

6a 0.04　**b** 0.16　**c** 0.28　**d** 0.34
7a 22, 23, 24, 30, 31, 32　**b** 7　**c** 41
8a It uses the binary system (powers of 2)　**b** 12p

PAGE 21 EXERCISE 3, 4F

1a 16.33　**b** 3.77
2 Astrid 2 min 52.98 s, Leila 2 min 53.34 s
3a 142 m, 67.2 m　**b** 74.8 m
4 £4.72
5a 3 min 35.8 s　**b** 1.8 s
6a Charles 20 m, Helen 19.6 m, Pat 19.25 m
b C–H 0.4 m, C–P 0.75 m, H–P 0.35 m　**c** F
7 Yes, by 2.68 kg

PAGE 22 EXERCISE 5, 6F

1a 58.2, 496, 1060; 1614.2　**b** 4.1, 9, 1250, 850;
2113.1
2a £973　**b** £2919　**c** £508　**d** £2540
e £292.50　**f** £1755　**g** £212　**h** £5300
3 40 × £13; £492.48
4 50 × 40 = £20, 20 × 100 = £20, 100 × 40 = £40,
30 × £1 = £30
5 £18.72, £19.44, £39.90, £35.96, £114.02
6a 80, 74.48　**b** 8000, 7715.4　**c** 36, 36.702
d 10 000, 10 291.9
7 513.9
8 5 min 37.5 s
9 £2550, £1912.50
10a 30 min, £7.83　**b** 26.1p

PAGE 23 EXERCISE 7F

1a 3.468　**b** 12.061　**c** 1.236　**d** 0.047
2a £8.38　**b** £2.02　**c** £1.10　**d** £0.81
3a 21 days　**b** 31 s　**c** 36.8°　**d** 4 tonnes
e 14.39 km　**f** 1.781 kg
4a and **d**
5a 15.364, 15.355　**b** 0.084, 0.075　**c** 7.034, 7.025
d 34.794, 34.785
6 £2.85, 5p
7 Kris, by 21p per hour
8 8.4
9a £1.97　**b** £2　**10** 13.9　**11** 7
12 1.25 kg
13 Sharon's by over 7p an hour
14a 27　**b** 1.4 kg
15 44

5 FACTS, FIGURES AND GRAPHS

PAGE 25 EXERCISE 1F

1a Red 14, black 13　**b** Hearts 7, Clubs 7, Diamonds
7, Spades 6　**c** A to 4, 8; 5 to 9, 9; 10 to K, 10
2a 5, 10, 6, 4, 3 families　**b** 28　**c** 2
3a 1, 2, 2, 3, 3, 3, 3, 5, 3, 3; 2, 2, 7, 4, 4, 1, 4, 3, 1
b 74, third
4a 3, 5, 10, 5, 2, 4　**b** They would be too spread
out　**c** 16–20
5a Broth　**b** 50　**c** Tomato 120, Chicken 170,
Broth 50, Leek 160　**6b** 80
7a 1992　**b** 1990–1991　**c** 1992, 9300; 1990, 7500;
1991, 6700; 1989, 6000　**d** Twelve symbols and one
with seven branches

PAGE 27 EXERCISE 2F

1a 30 000　**b** May　**c** (i) 20 000　(ii) 75 000
2a Increase until the third month, then fall
b (i) 3　(ii) 40 000　**c** 3 and 4　**d** 149 000
3a UK 56 m, Italy 54 m, France 52 m, Spain 36 m,
Belgium 10 m, Portugal 8 m　**b** 4
5a To reduce the spread
b 50–59　**c** 40　**d** 118
6a 0p–99p, 12; £1.00–1.99, 11; £2.00–2.99, 8;
£3.00–3.99, 4; £4.00–4.99, 5
7 17, 17, 21, 18, 18, 15, 13, 12, 22, 18
8a Cara　**b** Mabel £1.13, Karen 60p, Cara 33p
c £1.12, 1.13, 0.75; £1.60, 0.60, 1.40; £3.34, 0.33, 0.33
9a Books　**b** £600　**c** £200　**d** £200
10a March, April, May, December　**b** June,
September
c Mike 67 hours, Francine 60 hours
11 Approximate answers　**a** 3.5　**b** 165°C　**c** 67

6 MEASURING TIME AND TEMPERATURE

PAGE 30 EXERCISE 1F

1a 224　**b** Wed
2a 3600　**b** 1440　**c** 86 400　**d** 168　**e** 10 080
f 8760 or 8784
3a Thursday　**b** Tuesday or Wednesday if a leap
year
4a 3　**b** 19　**c** 21　**d** 25

PAGE 31 EXERCISE 2F

1 1.07 pm　**2** 12 min 20 s
3a (i) 18.5 s　(ii) 6.2 s　(iii) 47.7 s　**b** 14.6 s
4 15 13　**5** 11 h 48 min and 7 h 5 min
6 3 h 11 min 33 s　**7** 70 h 5 min　**8** 12 34 56; 7.8.90

PAGE 32 EXERCISE 3F

1a (i) A No. 62 bus to Galashiels and change to a No. 95
(ii) A No. 95 to Galashiels and change to a No. 62
b (i) Galashiels   (ii) 7 min, or 1 h 7 min, or 2 h 7 min
(iii) 11 31
2a 10 55   **b** No. 95 from Selkirk at 10 55, arriving Galashiels at 11 10. No. 62 from Galashiels at 11 12. (2 min for the change). Arrive Peebles at noon. Total time 1 h 5 min
3a No. 62 at 10 57, arriving Galashiels at 11 05. No. 95 from Galashiels at 11 15, arriving Edinburgh at 12 40
b 15 min
4 *First way.* A No. 62 at 11 50, arriving Edinburgh at 14 07; 2 h 17 min journey. *Second way.* A No. 62 at 11 50. Change at Galashiels. A No. 95 at 12 15, arriving Edinburgh at 13 40; 1 h 50 min journey. The first way avoids changing buses, but takes longer and ends with only 8 minutes to spare. The second way is shorter, arriving sooner, but requires a change of buses.

PAGE 33 EXERCISE 4F

1a (i) START—Subtract 32—Multiply by 5—Divide by 9—Add 273—STOP
(ii) START—Multiply by 9—Divide by 5—Add 32—STOP
(iii) START—subtract 273—Multiply by 9—Divide by 5—Add 32—STOP
b 572, 300, 573; 932, 500, 773; 1202, 650, 923; 1832, 1000, 1273
2a Diane 16°C; Jason 27°F   **b** He is using smaller units (°F)   **c** 20, 35, 50, 65, 80, 95 (all °C). His rise of 15°C is not as good as Diane's 16°C   **d** 20°C or 68°F, the starting time

PAGE 34 EXERCISE 5F

1a 2°, 0°   **b** 5°, −5°   **c** 11°, 9°   **d** −3°, −9°
2a 12°   **b** 7°   **c** 20°
3a −3°C   **b** (i) 9 am   (ii) 5 am and 7 am
c (i) −4°C   (ii) 6 am   **d** 4 hours   **e** 2, 2, 0, −3, −4, −3, 0, 4   **f** −0.25°C
4a −2°   **b** −8°   **c** −11°   **d** 0°   **e** −12°
5a −40°C   **b** −17.8°.

 COORDINATES: X MARKS THE SPOT

PAGE 35 EXERCISE 1F

1a C (−3, −2)   **b** B (4, −2), H (−1, −2)
c K (−3, −3)   **d** D (−4, 0)   **e** E (0, 3), F (2, 0), J (0, −4)   **f** F (2,0) and B (4, −2)
2a A (1, 2), B (4, 0), C (−2, −3), D (−3, 0), E (−2, 3), F (2, −2), G (−3, 1), H (1, −2), J (3, −4)

b (i) A, B, D   (ii) E, F, G, J   (iii) C, H
c (i) J   (ii) D. The faster they fly, the farther they have travelled (in straight lines)

PAGE 36 EXERCISE 2F

1a (−4, 9), (−5, 9), (−6, 10), (−6, 11), (−5, 12), (−4, 12), (−3, 11), (−3, 10)
b (±2, 5½), (±1, 5½), (±1, 6), (±4, 6), (±4, 7), (±5, 7), (±5, 5), (±2, 5), (±2, 6)
c Letter F and its image
2a The letter **h**: (10, −2), (9, −3), (8, −3), (7, −2), (7, 4), (8, 3), (8, −2), (9, −3), then (8, 1), (10, 3), (11, 2), (11, −2), (12, −3), (11, −3), (10, −2), (10, −3)
The letter **s**: (12, −3), (13, −3), (13, −1), (12, 1), (12, 2), (13, 3), (14, 3), (15, 2), (14, 3), (13, 2), (13, 1), (14, −1), (14, −2), (13, −3)
c The word is 'maths' when viewed normally, or upside down

PAGE 37 EXERCISE 3F

1a (11, 1), (15, 1), (19, 1)   (i) Values of x differ by 4, or x is a multiple of 4, less 1   (ii) y = 1
2a (2, 3), (6, 3)   **b** (5, 0), (5, 2), (5, 4), (5, 6)
c (1, 5), (5, 5)
3 From (16, 0) it runs along x = 16 to (16, 2), then along y = 2 to (9, 2), then along x = 9 to (9, 4), along y = 4 to (12, 4), along x = 12 to (12, 6), along y = 6 to (14, 6), along x = 14 to top of wall at (14, 8)
4 *Examples*   **a** (1, 1), (6, 6)   **b** (1, 3), (2, 6)
c (1, 5), (2, 10)   **d** (1, 3), (2, 4)   **e** (2, 0), (3, 1)
f (1, 5), (2, 8)

 SOLVING EQUATIONS

PAGE 38 EXERCISE 1F

1a 16   **b** 5   **c** 10   **d** 6   **e** 4   **f** 3   **g** 0
h 7   **i** 4   **j** 0   **k** 10   **l** 44   **m** 5   **n** 3
o 12   **p** 8   **q** 1   **r** ½   **s** ⅓   **t** 1½
2a 2 × 9 = 18   **b** 5 × 5 = 25   **c** 4 + 4 = 8
d 6 × 3 = 18   **e** 4 × 2 = 8   **f** 8 − 6 = 2
3a 3   **b** 5   **c** 7   **d** 9   **e** 3   **f** 9   **g** 2
h 12   **i** 23   **j** 42   **k** 51   **l** 9   **m** 2
n 4   **o** 83   **p** 13
4a 70   **b** 14   **c** 4   **d** 2

PAGE 39 EXERCISE 2F

1a 4   **b** 3   **c** 4   **d** 3   **e** 1   **f** 3   **g** 0
h 5   **i** 3   **j** 2   **k** 2   **l** 2   **m** 8   **n** 6
o 7   **p** 2   **q** 3   **r** 1   **s** 1   **t** 7   **u** 3
2a 5x + 4 = 14, x = 2; 4, 10   **b** 2x + 5 = 9, x = 2; 4, 5
c 4x + 3 = 19, x = 4; 5, 14

3a $12x + 9 = 27$, $x = 1\frac{1}{2}$; 3, 6, 9 **b** $7x + 5 = 866$, $x = 123$; 125, 126, 246, 369

c $11x + 4 = 279$, $x = 25$; 27, 30, 49, 73, 100

4 $4x + 30 = 270$, $x = 60$; 60 ml

5a $8x + 40 = 1000$, $x = 120$; 120 ml **b** 440 ml, 560 ml.

 9 MEASURING LENGTH

PAGE 41 EXERCISE 1F

5 4 cm, 8.6 m, 10.9 km.

PAGE 42 EXERCISE 2F

1a No **b** No **c** 120 km **d** Point where circles touch **f** Area common to both circles
h (i) 50 km (ii) 70 km (iii) 100 km
2 10.54 m, 10.46 m

PAGE 43 EXERCISE 3F

1a *Softwood* 1.8 m, 2.1 m, 2.4 m. *Hardwood* 1.8 m, 1.9 m, 2.0 m **b** 5.7 m, 6.0 m, 6.3 m **c** 16 **d** 1.8 m, 2.1 m, 2.4 m
2 10 **3a** 2m **b** 2
4a Lounge 7, 3; bedroom 9, 4; hall 10, 5; kitchen 7, 3; bathroom 5, 2 **b** £173.

 10 TILING AND SYMMETRY

PAGE 44 EXERCISE 1F

In some answers, different diagrams are possible

1a **b**

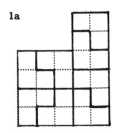
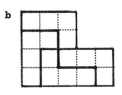

2

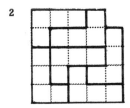

3a **b**

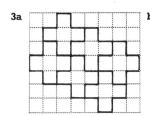
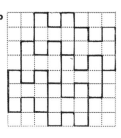

c

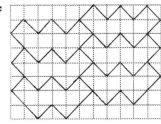

4a $(2, 3)$, $(4, 4)$, $(6, 5)$, $(8, 6)$, $(10, 7)$, $(12, 8)$, $(20, 12)$, $(2n, n+2)$ **b** $(0, 0)$, $(2, 1)$, $(4, 2)$, $(6, 3)$, $(8, 4)$, $(10, 5)$, $(18, 9)$, $(2n-2, n-1)$

5a

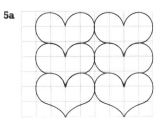

b

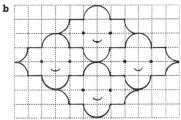

PAGE 45 EXERCISE 2F

1a 2 **b** 0 **c** 2 **d** 0 **e** 2 **f** 1 **g** 0 **h** 3
2a **b**

c **d**

3a **b**

c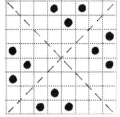

4a $(4, 6), (6, 5), (1, 4), (7, 3), (3, 2), (5, 2), (7 - b, 7 - a)$
b $(0, 0), (3, 4), (4, 5), (6, 2), (7, 1), (1, 3), (a, 6 - b)$
c (i) $(a, 8 - b)$ (ii) $(a, 10 - b)$ (iii) $(a, 2n - b)$
d $(2n - a, b)$

5a 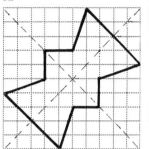 **b**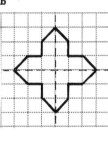

c 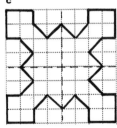 **d**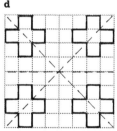

e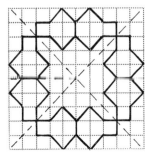

PAGE 47 EXERCISE 3F

1a Half **b** half **c** quarter **d** half
e quarter **f** quarter **g** half **h** quarter
i neither **j** quarter **k** half **l** quarter
m half **n** half **o** neither
2a 2 **b** 2 **c** 4 **d** 2 **e** 4 **f** 4 **g** 2
h 4 **i** 5 **j** 4 **k** 2 **l** 4 **m** 2 **n** 2
o 3

4a (i) 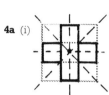 (ii)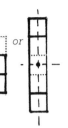

b (i) (ii)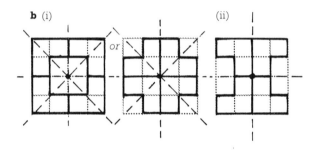

c (i) (ii)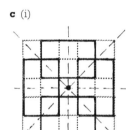
etc.

5a  **b**

c
etc.

11 MEASURING AREA

PAGE 48 EXERCISE 1F

1 26 cm²   **2** 22 cm²   **3** 200 m²   **4** 156 cm²
5 7100 cm²   **6** 15 m²   **7** 1400 cm²   **8** 504 cm²

PAGE 49 EXERCISE 2F

1 (*Rows across*) 36 m², £9, £324; 20 m², £17, £340; 27 m²,
£25, £675; 12 m², £12, £144; 21 m², £12, £252; 4 m², £9,
£36; 60 m², £25, £1500. TOTAL £3271
2a 240 cm²   **b** £1.58

PAGE 50 EXERCISE 3F

1a $67\frac{1}{2}$ cm²   **b** 48 m²   **c** 114 cm²   **d** 119 mm²
2a 6.75   **b** 17   **c** 24
3a 8   **b** 6   **c** $7\frac{1}{2}$ **d** $5\frac{1}{2}$—all sq. units
4a 18.2 m   **b** 2.88 cm   **5a** 2.4 km   **b** 445 m²

12 LETTERS, NUMBERS AND SEQUENCES

PAGE 52 EXERCISE 1F

1 (*Rows across*) **a** 3; 5; 7, 13   **b** 6, 10; 16, 2, 14
c 10; 40; 5, 15
2 60; 35, 14b; 30, 10, a
3 4; 20, 6, 42
4a (i) $P = a + b + c$ (ii) 41 cm
b (i) $P = s + 2t + 2u$ (ii) 60 cm   **c** (i) $P = 4k$
(ii) 50 cm   **d** (i) $P = 4p + 2r$ (ii) 640 cm
5a $L = 5x + 7y$ **b** 580 cm
6a $p + q + r$, 12, 16; 6; pqr, 48, 0, 36
b 6, 3; 2, 0; 5, 4, 3; 3, 20, 36, 9;
11, 23, 30, 20, 15; 9, 22, 30, 20

PAGE 53 EXERCISE 2F

1a 1000, 2500, 100n **b** 110, 220, 11n
c 80, 240, 8n **d** 150, 375, 15n **e** 250, 2500, 25n
2 6, 8, 10, 12, 20, 2n; 9, 12, 15, 18, 30, 3n; 5, 6, 7, 8, 12,
$n + 2$
3a 11, 21, $n + 1$ **b** 20, 25, $n + 10$
c 30, 45, $n + 20$ **d** 25, 50, $n + 15$ **e** 39, 56, $n + 29$
4b 4, 5, 6, 7, 11, $n + 1$; 6, 8, 10, 12, 20, 2n; 4, 5, 6, 7,
11, $n + 1$
5a 3, 3, 2, 1; 9   **b** (i) 4, 4, 3, 2, 1; 14   (ii) 5, 5, 4, 3,
2, 1; 20   (iii) $n - 3$, $n - 3$, $n - 4, \ldots, 3, 2, 1$

PAGE 55 EXERCISE 3F

1a (i) 90, 9n (ii) 16   **b** 120, 12n (ii) 12
c (i) 180, 18n (ii) 8   **d** (i) 25, $n + 15$ (ii) 129
e (i) 30, $n + 20$ (ii) 124
2a (i) 28, 3$n - 2$ (ii) 14
b (i) 37, 4$n - 3$ (ii) 16   **c** (i) 23, 2$n + 3$ (ii) 39
d (i) 62, 6$n + 2$ (ii) 17   **e** (i) 57, 5$n + 7$ (ii) 14
3 6, 10, 14, 18, 42, 4$n + 2$
4 5, 8, 11, 14, 32, 3$n + 2$
5 12, 20, 28, 36, 84, 8$n + 4$
6b 502 cm   **c** 17
7a 122 cm   **b** 86

13 TWO DIMENSIONS: RECTANGLE AND SQUARE

PAGE 57 EXERCISE 1F

1b All the horizontal lines are parallel and all the
vertical lines are parallel. The four drawers are
congruent rectangles, as are two cupboard doors
4a $y + 4x$ km   **b** $3y + 4x$ km   **c** $5y + 4x$ km
5 78, 36, $13\frac{1}{2}$, 42 feet
6 Four: (5, 4), (9, 4), (9, 7), (5, 7); (5, 4), (1, 4), (1, 7),
(5, 7); (5, 4), (1, 4), (1, 1), (5, 1); (5, 4), (9, 4), (9, 1), (5, 1)
7 16
8a (i) 1, 4, 9   (ii) 0, 1, 4   (iii) 0, 0, 1   **b** (i) 16, 9, 4
(ii) n^2, $(n - 1)^2$, $(n - 2)^2$

PAGE 58 EXERCISE 2F

1a 24   **b** 11
2a 13 right angles   **b** 6 angles of 56° and 6 of 34°
3a (i) 2, 4, 8   (ii) 16   (iii) $2^{10} = 1024$
(iv) $2 \times 2 \times 2 \times \ldots n$ times, or 2^n **b** (i) 4, 16, 64
(ii) 256   (iii) $4 \times 4 \times 4 \times \ldots 10$ times $= 1\,048\,576$
(iv) 4^n

PAGE 59 EXERCISE 3F

1 152 m   **2a** 5 units
3a Angles at corners all 20° and 70°; angles at centre
40° and 140°   **b** Angles at corners 32° and 58°;
angles at centre 116° and 64°
4a 40 cm   **b** 40 cm
5a 116 m   **b** (i) 9 m   (ii) 16 m   (iii) 14 m   (iv) 36 m
6 144 cm   **7** 10 m

PAGE 60 EXERCISE 4F

1a *Example:*   MOVE (0, 2)
                    DRAW (4, 2)
  b (2, 5)   DRAW (4, 8)
                    DRAW (0, 8)
                    DRAW (0, 2)

2a $(1, 2)$, $(6, 5)$

  b *Example*:   MOVE $(1, 5)$
                    DRAW $(6, 2)$
                    MOVE $(6, 5)$
                    DRAW $(1, 2)$

3 $(6, 3)$ should be $(3, 6)$

4 *Example*:   MOVE $(1, 2)$
                    DRAW $(5, 2)$
                    DRAW $(5, 4)$
                    DRAW $(1, 4)$
                    DRAW $(1, 2)$
                    DRAW $(5, 4)$
                    MOVE $(1, 4)$
                    DRAW $(5, 2)$

PAGE 61 EXERCISE 5F

1 Mark opposite sides equal and parallel, and all angles 90°. Make all the sides equal

2 $(5, 2)$

3 $(0, 4)$

4 E $(4, 3)$, F $(2, 5)$ and E $(8, 7)$, F $(6, 9)$

5 $(-7, -6)$, $(-7, 4)$, $(3, 4)$, $(3, -6)$

6 The diagonals bisect each other. Make them at right angles to each other.

7a Sides 10, half-diagonals 7   **b** Angles at corners 45° and 90°, angles at centre 90°

8 $(3, 5)$, $(7, 1)$

9 All sides equal and all angles 90°; diagonals are equal and bisect at right angles; symmetrical about diagonals and lines through midpoints of sides

10 $(1, 4)$, $(7, -2)$, $(1, -8)$, $(-5, -2)$

11a At centre of square; 6 cm   **b** Draw perpendicular lines through centre for the diagonals; 11.3 cm

12a A $(3, 1)$, C $(8, 6)$, E $(13, 11)$, G $(18, 16)$, I $(23, 21)$
  b (i) $(5n - 2, 5n - 4)$ (ii) $(498, 496)$

PAGE 62 INVESTIGATION

a (i) 5   (ii) 7   (iii) 9   (iv) 13   **b** 13, 15, 17, 21

PAGE 62 EXERCISE 6F

1b MOVE $(0,0) \ldots (4,0) \ldots (6,2) \ldots (6,4) \ldots (5,5) \ldots (4,5)$
    DRAW $(8,0) \ldots (8,4) \ldots (6,6) \ldots (4,6) \ldots (3,5) \ldots (3,4)$
    DRAW $(8,8) \ldots (4,8) \ldots (2,6) \ldots (2,4) \ldots (3,3) \ldots (4,3)$
    DRAW $(0,8) \ldots (0,4) \ldots (2,2) \ldots (4,2) \ldots (5,3) \ldots (5,4)$
    DRAW $(0,0) \ldots (4,0) \ldots (6,2) \ldots (6,4) \ldots (5,5) \ldots (4,5)$

2 *Examples*

a MOVE $(4, 4) \ldots$ DRAW $(2, 4) \ldots (1, 2) \ldots (4, 4) \ldots$
  $(5, 2) \ldots (1, 2) \ldots (2, 1) \ldots (4, 1) \ldots (5, 2)$

b MOVE $(0, 0) \ldots$ DRAW $(6, 0) \ldots (3, 1) \ldots (0, 0) \ldots$
  $(0, 2) \ldots (3, 1) \ldots (6, 2) \ldots (0, 2) \ldots (3, 4) \ldots$
  $(6, 2) \ldots (6, 0)$

c MOVE $(2, 6) \ldots$ DRAW $(0, 3) \ldots (2, 3) \ldots (2, 6) \ldots$
  $(5, 3) \ldots (2, 3) \ldots (2, 2) \ldots (0, 3) \ldots (2, 0) \ldots$
  $(2, 2) \ldots (5, 3) \ldots (2.0)$

---

14 MEASURING VOLUME

PAGE 63 EXERCISE 1F

1a 96 cm³   **b** 65 cm³

2a 192 000 cm³   **b** 872 000 cm³   **c** 72 000 cm³

3 345 600 cm³

4 93 cm³

PAGE 64 EXERCISE 2F

1a (i) 64 cm³   (ii) 8 cm   **b** 2 cm × 4 cm × 8 cm, 64 cm × 1 cm × 1 cm, etc.

2a 3 cm   **b** 3 cm

3a (i) 26   (ii) 7   (iii) 4   (iv) 3

  b 8 cm × 8 cm × 8 cm, and larger cubes.

---

15 FRACTIONS AND PERCENTAGES

PAGE 65 EXERCISE 1F

1a As a record speed   **b** in music
  c door, window, eyes, etc.   **d** diving, aerobatics, etc.

2a $\frac{5}{26}$ **b** $\frac{21}{26}$

3 $\frac{8}{12}$ or $\frac{2}{3}$

4 John $\frac{1}{7}$, Jenny $\frac{2}{7}$, Jim $\frac{4}{7}$

5

Fold in half   Fold over   Unfold

6a 10p   **b** 50p   **c** 20p   **d** 1p   **e** 2p   **f** 5p

7a A degree   **b** decilitre   **c** milligram

8a (i) Divide each pie into fifths; each person gets 4 parts

(ii) Divide each pie into fourths; each person gets 5 parts

  b (i) $\frac{4}{5}$ (ii) $1\frac{1}{4}$

9a

3 triangles each
  b $\frac{3}{4}$

10

 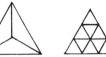

11a 0.667   **b** 0.8   **c** 0.317   **d** 1.167   **e** 3.217
  f 5.983

12a $\frac{8}{9}$ **b** $\frac{3}{11}$ **c** $\frac{2}{15}$

PAGE 66 EXERCISE 2F

1a $\frac{4}{7}$ **b** $\frac{5}{6}$ **c** $\frac{3}{8}$ **d** $\frac{2}{3}$ **e** $\frac{4}{5}$ **f** $\frac{3}{10}$
2a $\frac{7}{10}$ **b** $\frac{1}{4}$ **c** $\frac{1}{20}$ **d** $\frac{6}{25}$ **e** $\frac{19}{20}$ **f** $\frac{21}{50}$
3a $\frac{7}{15}$ **b** $\frac{8}{15}$ **c** $\frac{1}{3}$ **d** $\frac{1}{5}$ **e** $\frac{2}{25}$
4a $\frac{1}{4}$ **b** $\frac{1}{3}$ **c** $\frac{2}{9}$ **d** $\frac{5}{6}$ **e** $\frac{5}{9}$
5a $\frac{3}{4}$ **b** $\frac{1}{8}$ **c** $1\frac{1}{20}$ **d** $2\frac{9}{25}$

PAGE 66 EXERCISE 3F

1a 100 **b** $166\frac{2}{3}$
2a 60, 30, 15 **b** each is half of the previous one
3 $\frac{5}{6}$ of £30 **4** 13, $32\frac{1}{2}$ and 117 miles
5 Smallest $12\frac{3}{5}$ by $2\frac{7}{10}$ by $\frac{4}{5}$in, largest $13\frac{3}{10}$ by $2\frac{4}{5}$ by $1\frac{1}{5}$in

PAGE 67 EXERCISE 4F

1a 50% **b** (i) Twice as much (ii) 4 times as much
(iii) 20 times as much (iv) 10 times as much
c (i) £4 (ii) £8 (iii) 40p (iv) £20
2a 18%, 0.18, $\frac{18}{100}$, $\frac{9}{50}$ **b** 35%, 0.35, $\frac{35}{100}$, $\frac{7}{20}$
c 6%, 0.06, $\frac{6}{100}$, $\frac{3}{50}$ **d** 52%, 0.52, $\frac{52}{100}$, $\frac{13}{25}$
3 70%, 0.7, $\frac{70}{100}$, $\frac{7}{10}$; 2%, 0.02, $\frac{2}{100}$, $\frac{1}{50}$; $12\frac{1}{2}$%, 0.125,
$\frac{12.5}{100}$, $\frac{1}{8}$. The missing piece is $\frac{40}{100}$, to make 40%, 0.4,
$\frac{40}{100}$, $\frac{2}{5}$ **4a** 48.3% **b** 51.7%

PAGE 68 EXERCISE 5F

1a £43.75, £293.74 **b** £293.74
c Yes; 100% + 17.5% = 1.175 in decimal form
2a $\frac{11}{7}$ of £12.88 **b** By 8p
3a $\frac{7}{100}$ **b** 7%
4a 8.2% **b** £73.80, £85.28, £53.36, £50.60
5a £1438.33 **b** £10068.31
6a (i) 76%A, 52%C (ii) 32 or over **b** 16 or over
7a 28% **b** 2.8kg, 0.5kg, 0.8kg **c** A, 6%
(above); B, 4.6% (below); C, 4.4% (below); D, 5.6%
(above)

16 SOLVING MORE EQUATIONS

PAGE 69 EXERCISE 1F

1a 3 **b** 0 **c** 10 **d** 2 **e** 4 **f** 5 **g** 3 **h** 4
2a $3x$ **b** $4x$ **c** $5x$ **d** $2x$ **e** $6a$ **f** b
g $3c$ **h** d **i** $4n$ **j** $5n$ **k** $2n$ **l** 0
3a $7t = 14$, $t = 2$ **b** 3 **c** 1 **d** 1 **e** 7
f 3 **g** 4 **h** 5 **i** 4
4a 1 **b** 5 **c** 5 **d** 7 **e** 2 **f** 1 **g** 8 **h** 8

PAGE 69 EXERCISE 2F

1a $11x$ **b** x **c** $16y$ **d** $6y$ **e** 0 **f** 1
g 3 **h** $4t - 1$ **i** $5s + 3$ **j** 3

2a $\begin{array}{r} 2x + 5 \\ -2x \\ \hline 5 \end{array}$ **b** $\begin{array}{r} 8x + 1 \\ -8x \\ \hline 1 \end{array}$ **c** $\begin{array}{r} 3 - 2y \\ +2y \\ \hline 3 \end{array}$ **d** $\begin{array}{r} 4 - m \\ + m \\ \hline 4 \end{array}$
e $\begin{array}{r} m + 4 \\ -m \\ \hline 4 \end{array}$ **f** $\begin{array}{r} 10 - t \\ + t \\ \hline 10 \end{array}$ **g** $\begin{array}{r} 4k + 4 \\ -4k \\ \hline 4 \end{array}$
h $\begin{array}{r} 1 - x \\ + x \\ \hline 1 \end{array}$ **i** $\begin{array}{r} 13a + 9 \\ -13a \\ \hline 9 \end{array}$ **j** $\begin{array}{r} 15 - 13k \\ + 13k \\ \hline 15 \end{array}$

3a 5 **b** 1 **c** 0 **d** 0 **e** 3
4a (i) $2x + 3$ (ii) $4x + 3$ **b** (i) $3y + 4$ (ii) 4
c (i) $2t + 7$ (ii) $t + 7$ **d** (i) $3u + 5$ (ii) $4u + 5$
e (i) a (ii) 0 **f** (i) $2c - 2$ (ii) $3c - 2$
g (i) $3d - 6$ (ii) $d - 6$ **h** (i) $2x - 6$ (ii) x

PAGE 70 EXERCISE 3F

1a 4 **b** 1 **c** 2 **d** 1 **e** 5 **f** 1 **g** 5 **h** 12
2a $3x = x + 6$, $x = 3$ **b** $3x = 2x + 1$, $x = 1$
c $5x = 2x + 3$, $x = 1$ **d** $4a = 2a + 6$, $a = 3$
e $3r = r + 6$, $r = 3$ **f** $7a = 5a + 10$, $a = 5$
g $2b = b + 8$, $b = 8$ **h** $5a = 2a + 6$, $a = 2$
i $2c + 5 = c + 13$, $c = 8$
3a 10th **b** 15th **c** 4th
4a 3 **b** 4 **c** 17 **d** 9 **e** 25 **f** 14
g 3 **h** 12 **i** 25 **j** 13 **k** 2 **l** 2

PAGE 71 EXERCISE 4F

1a 7 **b** 3 **c** 7 **d** 2 **e** 9 **f** 1 **g** 4 **h** 2
2a $2x + 5 - 3 = x + 4$, $x = 2$
b $4x - 5 = 3x + 2$, $x = 7$ **c** $2a + 4 = 12 - 2a$, $a = 2$
d $17 - 2c = 20 - 3c$, $c = 3$
e $5 - 2n + n = 2n + 2$, $n = 1$
f $2x + 1 - 2 = 5 - x$, $x = 2$
3a $3h = 2h + 50$, $h = 50$, £1.50
b $4s = 2s + 10 + 22$, $s = 16$, 64p
c $5a + 10 = 3a + 52$, $a = 21$, £1.15
d $2c + 3c + 15 + 40 = 4c + 5c + 35$, $c = 5$, 80p

17 THREE DIMENSIONS

PAGE 73 EXERCISE 1F

1 b, e, g, i, j, k
2a 6 faces, 8 corners, 12 edges **b** 12 edges
c 8 corners **d** 12 edges **e** 6 faces **f** 6 faces
g 8 corners
3 All edges equal, all faces square, fits a close-fitting
box in more than 4 ways

PAGE 75 EXERCISE 3, 4F

1a (i), (ii), (iv); the others have card left over which the machine can't deal with
b (ii), (iii), (v)
c (i) Method 1

L	28	26	24	22	20	18	16	14	12
B	1	2	3	4	5	6	7	8	9
H	9	8	7	6	5	4	3	2	1
Sfc. area	578	552	522	488	450	408	362	312	258

Method 2

L	18	16	14	12	10	8
B	1	2	3	4	5	6
H	14	13	12	11	10	9
Sfc. area	568	532	492	448	400	348

(ii) $12 \times 9 \times 1$ from Method 1
2a 4 **b** (i) 4-way fitting, half-turn symmetry, opposite sides equal, angles 90°, etc.
(ii) 4-way fitting **c** (i) 8 (ii) 4 **d** (i) $8 \times 4 = 32$
(ii) Pair AH (then 2H, 3H, 4H) with each club card and AC (then 2C, 3C, 4C) with each heart card
e 384, 6144

PAGE 76 EXERCISE 5F

1 30 units
2a 3 over; **b** 5 over; **c** 2 over; **e** 6 over; **f** 7 over;
h 0 over
3b Any model in which two of length, breadth, height add up to 15 units and the third is 10 units or less

PAGE 77 EXERCISE 6F

1a
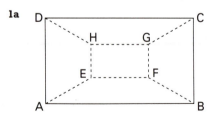

b (i) EFGH (ii) CDHG (iii) ADHE
c Top left corner of back wall, or top right corner of left wall, or far left corner of ceiling
2a 8 **b** 2 **c** AB, PQ, RS, CD; AQ, BP, CS, DR; AD, QR, PS, BC. Cuboid has base DCSR, top ABPQ.

18 PROBABILITY

PAGE 78 EXERCISE 1F

2 Even chance
3 Even chance
4 Unlikely
5a Even chance **b** Even chance **c** Unlikely
d Likely
6 Impossible (the speed of light is much greater than the speed of sound)
7 Even chance
9 Certain
10 Even chance

PAGE 79 EXERCISE 2F

1a $\frac{1}{8}$ **b** $\frac{3}{8}$ **c** $\frac{1}{2}$ **d** $\frac{1}{2}$
2a $\frac{1}{8}$ **b** $\frac{3}{8}$ **c** $\frac{1}{4}$
3a $\frac{1}{3}$ **b** (i) $\frac{2}{3}$ (ii) $\frac{1}{6}$ (iii) $\frac{1}{6}$
c The form of all teams stays the same for every game, and each team has the same amount of good or bad luck
4 $\frac{9}{26}$
5a $\frac{3}{5}$ **b** $\frac{1}{5}$
6 $\frac{99}{991}$
7a (i) 2 (ii) $\frac{1}{2}$ **b** $\frac{1}{6}$
8a 6 **b** $\frac{1}{6}$
9a 6 **b** $\frac{5}{6}$